Effective Project Management

Effective Project Management

Guidance and Checklists for Engineering and Construction

Garth G.F. Ward

Registered Offices
John Wiley & Sons, Inc., 111 River Street, Hoboken, NJ 07030, USA
John Wiley & Sons Ltd, The Atrium, Southern Gate, Chichester, West Sussex, PO19 8SQ, UK

Editorial Office
9600 Garsington Road, Oxford, OX4 2DQ, UK

For details of our global editorial offices, customer services, and more information about Wiley products visit us at www.wiley.com.

Wiley also publishes its books in a variety of electronic formats and by print-on-demand. Some content that appears in standard print versions of this book may not be available in other formats.

Library of Congress Cataloging-in-Publication Data

Names: Ward, Garth G.F., author.
Title: Effective project management : guidance and checklists for engineering and construction / Garth G.F. Ward.
Description: Hoboken : Wiley-Blackwell, 2018. | Includes index. |
Identifiers: LCCN 2018005477 (print) | LCCN 2018010035 (ebook) | ISBN 9781119469421 (pdf) | ISBN 9781119469360 (epub) | ISBN 9781119469445 (paperback)
Subjects: LCSH: Project management. | Leadership. | BISAC: TECHNOLOGY & ENGINEERING / Construction / General.
Classification: LCC HD69.P75 (ebook) | LCC HD69.P75 W367 2018 (print) | DDC 620.0068/4–dc23
LC record available at https://lccn.loc.gov/2018005477

Cover design: Wiley
Cover image: © hh5800/iStock Photo;
© Pobytov/iStock Photo

Set in 10/12pt MyriadPro by SPi Global, Chennai, India

Printed in Singapore by C.O.S. Printers Pte Ltd

10 9 8 7 6 5 4 3 2 1

To Graham Ritchie who wrote the original list.
And to my wife Gwyneth who uses lists all the time.

Contents

Key to Figures: Figure: V-E-3
The initial Roman numeral is the book Part number
The middle Capital Letter is the Section of the book Part already indicated
The last Arabic numeral is the sequential figure number for the Section
concerned

Preface

When Graham Ritchie and I were project managers at Bechtel, Graham started writing a book on project management and invited me to be his co-author. Before Graham left to take up the post of director of the MSc in project management at Cranfield School of Management, we developed a large amount of material, probably sixty percent of the book. The reason this book has never been completed is that, after my sixteen years as a consultant lecturer at Cranfield, I realised that the material needed rewriting completely. Since my first book[1] only covered aspects of purchasing, it would probably take at least another two or three volumes to do justice to the whole subject of project management.

During Graham's short time at Cranfield, before he died suddenly of a brain tumour, he turned much of the material into lecture notes in support of the MSc programme. Naturally, when I took over from Graham, I continued to use them during my ten years as the MSc course director. Section A of Part one, Project Characteristics, Advantages and Phases has been revised and is included at the start of Part I of this book. One feature of our book was that each chapter would end with a check list covering the chapter topic. Accordingly, the check lists were also complied into a handout for the MSc students, and this forms the core of Part IV, Project Execution. This document was so useful for many of the graduates that, years later, they would write to me asking for another copy because theirs was worn out.

The popularity of the project execution check lists, and their practical application by the MSc graduates in the work environment, has been the reason for developing the original lists and taking the check list concept further.

The inclusion of the Specialist Topics is intended to extend the lists to cover subjects needed for projects of a significant size and complexity. Whereas, the Skills Check Lists have been included to cover the needs of individuals, in all circumstances, regardless of the size of the project.

---

1 The Project Manager's Guide to Purchasing, Contracting for Goods and Services by Garth Ward. Gower 2008.

Acknowledgements

I am particularly appreciative of the following people who have contributed to various sections. They responded to my requests for input, despite their busy lives, where their expertise was based on having done it for real. Others kindly gave permission to use existing published material. Some of the contributions have been used with little change and some have had content added. However, I trust that the integrity of the original material has been maintained. All their contributions have enhanced the knowhow of particular topics, broadened the subject base and added usefulness to the book as a whole.

Professor John Adair: over the years I have championed the John Adair Leadership Model application to project management and am appreciative of his giving permission to use his copyrighted Leadership Model.

Martin Arter MSc: Director of Infrastructure Programme Management at Network Rail brought clarity to the difference between project management and programme management by having done it in reality. His experience has involved managing both major projects, encompassing aerospace, power, and construction, and major programmes. Programmes he has managed have concerned the design and construction of highly complex navy warships and reorganizing track renewal at Network Rail into regions for effective programme management. His breadth of experience gives real authority to the subject.

Stephen Carver MSc: I am indebted to my friend and colleague for the concept and story outline of the air traffic control analogy, describing the management of portfolios. Listening to his charismatic and innovative presentations demonstrates that he is the person to demystify any subject.

Mike Cleaver MSc: who I always knew would achieve success in project management, is a vice president with CB&I. Mike, having been a manager of projects, as well as a project manager, has provided the essential information that the readers of this book need to be aware of when they are reporting to senior management.

I am especially indebted to **Cranfield School of Management** for permission to use the material that Graham developed for the project management MSc course. In particular, the subject matter that introduces Part I and also the original check lists that form the basis of Part IV on project execution and the foundation of this book.

Once more I must express my gratitude to **Vernon Evenson**, project manager, colleague and friend who was most helpful in providing fresh perspectives and new experiences. As well as validating my own know-how, he performed the invaluable task of giving me confidence in my own experiences. Vernon's contributions have all been integrated into the text where appropriate.

Pat McHugh: one time business unit leader in B.P., at the height of its successes, provided the client's perspective and the valuable section on Contracting Strategy Considerations. I have been an admirer of Pat's project management professionalism ever since we met on a short course for the state oil industry in Thailand many years ago.

Professor David R. Middleton: kindly gave me permission to use extracts from his book *Financial Decisions* for the section on Financial Appraisal. This book taught me all about financial appraisal and covers everything that a project manager will need to know about finance. It is published by Longman, fourth impression 1987, Understanding Business Series ISBN 0 582 35401 3.

I indebted to my son **Giles Ward,** who had his west-end film premier on the same hot summers day that England was playing Germany in the European cup, for explaining the differences in managing a theatre production and the making of a film. He is now a director of a building and construction company.

I must thank Nurse Practitioner **Jane A. Williams** for allowing me to cross examine her whilst cruising down the Volga (on our way from Moscow to St Petersburg) in order to identify the differences for a surgical project. This has been validated and added to from my own experience.

I must also, once again, thank **David Wright**. As a well-known beer advertisement said: "probably the best" exponent of contract law in the business. David has provided essential support on courses we have run together, and over the years he has allowed me the invaluable luxury of being able to pick his brains. He provided new material for the sections on contracts as well as expanding some of the material that Graham had developed.

It would be remiss of me not to thank all the lecturers who supported the Cranfield MSc course in Project Management. They were all pre-eminent in their field and shared valuable experiences. All these people provided answers to my many questions and provided me with an excellent education! They have all contributed in some way to this book.

I must not forget involvement from other family members: my son **Gavin Ward**, who is general manager of the UK office of a risk consultancy business. He has worked for a number of oil companies and has always been available to check facts and provide a client perspective.

Also **Dr Guy Ward,** who is the investment manager for a mega-million-pound private family investment trust, performed the invaluable task of reading the manuscript. Guy checked for readability and for those inconsistencies that an author can no longer identify after many iterations of the text.

Finally, I cannot overlook my granddaughter **Emily** who volunteered to do some proof reading and my wife **Gwyneth**, who was occasionally badgered into doing a little.

Introduction

When I joined the projects department of Shell Mex and BP, I was given a manual titled 'Policy Guide.' Naturally on the first project I was responsible for, I followed the policy religiously. Nevertheless, I was called into my boss's office and rapped over the knuckles and told: "For heaven's sake use your imagination – it's only a guide." Sometime later, on a major project, I had the vehicle maintenance facilities for road tankers redesigned. However, this had consequences for all the other similar projects that were being carried out at the same time. Not surprisingly I was called into the boss's office again, who banged the table and stated firmly "*That's the policy*!" I have always thought that this was brilliant. The skill of project management is deciding when to follow policy and when to do things differently. Consequently, this book is a *policy guide*. It may be good policy to follow the theoretical reasoning that is included, but the checklists will provide a guide to a practical approach.

This book is designed to help project managers achieve success. Its purpose is, to make project management 'boring.'[1] The thesis is that if your project is exciting, then it is in trouble. The book should help experienced project managers as well as those with less experience. Nevertheless, I shall never forget the one occasion (and the only one!) during my career as a practicing project manager when, to my surprise, I found that I had nothing to do. So, I decided to read the company project management manual. I was taken aback when I discovered something I had not thought about and which deserved investigation. I heaved a sigh of relief at my good luck, when the actions I took as a result prevented a project debacle. I trust that this book will provide the reader with some similar luck. Yet, there is no such thing as luck in project management. Luck comes from preparation and planning.

This is a guide to a broad spectrum of basic principles. Nevertheless, project managers also need to be conversant with today's rapidly changing technologies, particularly the information technologies. However, it is a failure to understand and conform to the basic principles that still cause projects to have problems. The knowledge required of a project manager is more than they can possibly know or know how to deliver on their own, and in any case we all make mistakes. Consequently, a checklist provides a system whereby the project manager can be confident that a valid process is being followed. This then enables them to be innovative in their approach to the project with the development of plans or resolution of problems.

---

1 I have borrowed this concept from my friend and colleague Stephen Carver.

Effective Project Management: Guidance and Checklists for Engineering and Construction,
First Edition. Garth G.F. Ward.
© 2018 John Wiley & Sons Ltd. Published 2018 by John Wiley & Sons Ltd.

I am aware that some people have difficulty with checklists. When the original loose-leaf paper organizers were all the rage, there was a sense of having somehow failed if you were not a dedicated user. Consequently, the book format has been chosen to provide people resistant to lists with many areas where the issues have expanded explanations. I am one of those who do not find using lists a natural process. On too many occasions, I would ask Gwyneth, my wife, to fax me some document to a remote part of the country because I had forgotten to take it with me. As a consequence, my son Guy has consistently given me a hard time, and rightly so, for not using a checklist when travelling on an assignment.

Apart from the early sections of Part I and Part II, which are of a more conventional book nature, the remaining structure is a pick-and-mix checklist format to enable the reader to select a flexible approach to those elements that they need. The first sections in Part I deal with some basic characteristics of projects and project management because the better these characteristics are understood, the more effective the management process will be. However, it is difficult to separate projects and project management. For example: the scope is very much part of a project, but it is a project management function to define the scope. One could say it is an objective of project management to complete the scope. Nevertheless, I have tried to separate them.

The book is intended to be highly practical and is based on experience. 'Experience is a truer guide than the words of others.'[2] However, to be truly effective, a project manager needs to be aware of the theory behind the issues concerned. Knowledge of the theory makes the subject interesting by observing how the theory works in practice. It also enables the reader to modify the advice given to suit different circumstances. Further, one can never be sure of people's level of expertise or what they know. For these reasons, brief elements of theory are included in the initial paragraphs of the various sections. It is designed to be as generic as possible and does not promulgate any particular method of working. Additionally, it must be remembered that the principles of project management are the same for all projects.

Rather than just produce a list of *what* activities need to be performed, the book also offers some advice on *how* tasks should be carried out. The challenge has been to cover all the essential elements needed by a project manager, in as concise a manner as possible, without compromising the issues under examination.

I hope that the reader can be like one of my MSc students who made sure that, regardless of how he felt or what he knew, he took away at least one grain of sand from every session, lecture, or assignment. He validated his knowledge, acquired a technique, or borrowed an experience. The result was that at the end of the course, he had a big pile of sand, and I would like to think that as a direct result, he went on to lead one of the major contractors.

The book is written primarily from the perspective of the project manager of the organization performing the work – the contractor. The term *client* is used to define the organization requesting the work. Accordingly, I have tried not to forget that a contractor performs the client role for their subcontractors and suppliers. In the text, substitute 'management approval' for 'client approval,' for different contract situations.

There is no set format to the individual sections. Each section varies according to the requirements of the topic. A few issues may appear in more than one list due

2 *Thoughts on Art and Life* by Leonardo Da Vinci.

to the structure of the lists. Some are more process-oriented, whereas others are more subject-focused. For example, 'obtain a project cost code' is part of the *project launch process*, but it is also part of the subject of *project control*. This emphasizes the importance of the issues concerned and ensures that they do not get overlooked. Where the same issue is discussed in a different context, a cross reference is provided.

Everything is described in a manual format on the basis that any computerised electronic system will still need to replicate a manual process. This book does not address the software that is available in the marketplace.

For completeness I have included a list of abbreviations in use within the project management world. Nevertheless, I trust that the terms I have used are self-evident in their own context. I have avoided academic debate over the meaning of terms,[3] and where I have used terms that are different from a norm, I have explained my reasons. For example, the Association for Project Management (APM) has specific definitions for projects, programmes, and portfolios, some of which one might wish to express differently. I could, for example, explain the term *programme* in two ways: firstly, to describe large projects such as the Olympics, Crossrail, or the space shuttle programme; and secondly, describing a smaller portfolio of miscellaneous business projects in an organization or an equipment maintenance programme. However, these explanations might not conform to the specifics of the APM definitions. Nevertheless, the important thing is that, in summary, the APM states 'the concept of projects, programmes and portfolios should be thought of as just points on a gradual scale of managing effort to deliver objectives.' Good, it is still project management!

---

3 See the Association of Project Management 'Body of Knowledge' for a Glossary of Terms.

PART I

Projects and Their Management

Section A   Project Characteristics and Phases[1]

Projects can be anything: a capital facility, an information system, a piece of research, a company merger, an organizational change, launching a product, or decommissioning a facility, and so on. They can range from capital intensive technological and infrastructure investments to labour-intensive health care. All projects types need a description, a scope, and the associated specifications for the quality required, and they cannot be realized without a team of people to develop them. The fundamental characteristic of all projects is that they create and cause change. As such, they come up against resistance. Consequently, leadership is needed in the form of a project manager, and a project management process is required to control them. (See Section B.)

There is a hierarchy to projects determined by their size, complexity, and the inherent risks (see Figure I.A.1). At the lowest hierarchical level are the *routines*, tasks that are so common and so well developed in a function that methods of working have ironed out all the difficulties. Next in the hierarchy are those frequently occurring packages of work – small projects that are very similar and can be developed without too much specialized management and theoretically do not present any significant risks. There are lots of them in an organization and they can be performed without any real difficulty. These are called the *runners*.

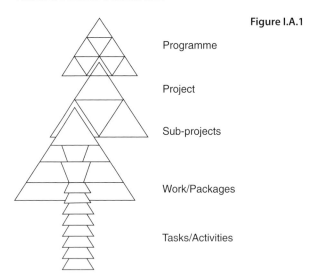

Figure I.A.1

Programme

Project

Sub-projects

Work/Packages

Tasks/Activities

At the next level of the hierarchy are larger projects that the organization performs reasonably regularly; they are very similar or replicate previous projects. Naturally, they are called *repeaters*. The development of repeaters has become more specialized, less routine, and more individually project-focused. As a consequence they have a higher risk of failure. They need someone experienced in project management because the real risk with them is that people assume they are repeats. The reality is that they have differences that, if ignored, could cause project failure. Then come the projects that

---

1 This section is based on Graham Ritchie's first Cranfield lecture note 'Project Management Characteristics and Advantages' and a second note 'Project Phases'.

are infrequent and more unusual, they become *strangers* to an organizations normal method of working. They are large projects and are high risk projects as far as the organization is concerned. As a consequence, they need someone to manage them, who is skilled and experienced in project management. Finally, the mega project, the first of a kind, the once-in-a-career opportunity are the *aliens,* consisting of a programme of large projects. (Part II addresses programme management).

'Every project begins on paper as an ideal, as a vision of perfection and quickly becomes mired in the confusion of budget, size and opposition of NIMBY's'.[2] Consequently, the best way to start a project is to carry out a feasibility study that results in a clear brief and statement of the requirements (see Part III). Nevertheless, there are features that are common to all projects regardless of size.

1 Characteristics

1.1 Unique Non-repetitive

A primary characteristic of projects: a product, a development, a task, or a deliverable is that they are unique and are non-repetitive.

1.2 Phases

Secondly, because projects start with a unique idea or concept, they go through a series of growth phases in order to achieve an outcome.

1.3 Risk

Projects are risky due to the very uniqueness of their nature. The risks are then compounded by the changes that can occur during the project's development. The severity, impact and consequences of the risks incurred are related to the hierarchical position of the project, as described above.

1.4 Business Objectives

Projects come into being because they will provide benefits to an owner and a return on their investment; they have a purpose. The business requirements of a project become the owner's objectives. They then get translated into specific objectives for the management of the project (see Section B).

1.5 Liable to Changes

Projects almost invariably change in scope, often by very large factors, due to changing business requirements and market conditions.

It is typically necessary to reduce the costs involved in order to make the project financially viable and to make the business case acceptable. This will usually mean reducing the scope or specification of the project – see Figure I.A.2. Everyone creating a project has big ideas, but when the budget can't get any bigger, the ideas have to get smaller.

---

2 Slightly adapted (project for 'House'), introduction to *House of the Year* programme 2016 by Kevin McCloud, TV Channel More 4.

Cost

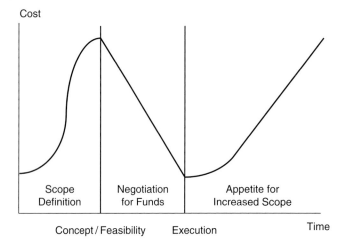

| Scope Definition | Negotiation for Funds | Appetite for Increased Scope |

Concept / Feasibility   Execution   Time

Figure I.A.2

1.5.1

Clients may have limited funds and reduce the budget, but they do not reduce their ambition. In reality, there is no such being as a client who does not make changes. Thus, once the project is approved and has got the go-ahead, there is a natural tendency to want to put back all the features that were removed. These changes are then likely to cause failure of the business objectives. Sometimes it can be almost impossible to match the requirements with the money that has been allocated to a project. This often occurs in the public sector. The correct approach is to deliver the essentials and, if there is anything left in the budget, add the 'nice to haves' when the essentials have been completed.

2 Phases

2.1

The development of a project is modelled by a series of phases or stages. Sometimes the phases they go through are carried out sequentially, but more often they overlap significantly. There are three basic project stages:

> Thinking – (Planning) – Doing

2.1.1

The conceptual, creative, thinking phase is a natural process; people enjoy it. The same is true of doing. People like to design, make, and construct – doing is also a natural process. This is not true of planning. Planning is not a natural process. It is imposed on a project's development by the management process. The trouble is that people's natural desire is to jump straight from the thinking stage to the doing stage. If this occurs, project disaster is guaranteed. It is not a flaw in the characteristics of projects; it is a failure in a project's management.

2.1.2

For more complex situations, these basic phases are broken down into more detail. Between each phase, there is an opportunity to assess the viability of the project and decide if one wishes to proceed to the next phase. The objective of breaking the project into phases is to enable one to plan and control the work at the appropriate level of detail.

2.2   Phase Details

The following Figure I.A.3 is a basic model of a typical project, showing the state of development for each phase of the project. These phases are typical for technological projects such as the process and power industries[3], but it is also intended to be generic:

a. Concept: a company, government, or some other body determines that there is a requirement for a new facility, plant, or product.
b. Feasibility: the concept is examined in detail to see whether it is a realistic, viable business proposition. This selection and definition phase is one of value creation.
c. Planning: if it is viable, an execution plan is developed.
d. Basic design: before major funds are committed, the basis of design is carefully agreed.
e. Design: once the basic 'recipe' is firm, detailed drawings for each element or component are produced.
f. Procurement: all the necessary services, materials, and equipment are purchased.
g. Construction or production: the facility or product is assembled from the materials and equipment, using the drawings already prepared.
h. Commissioning or setting to work: the plant or product is thoroughly tested to ensure that it satisfies the requirements of the project.

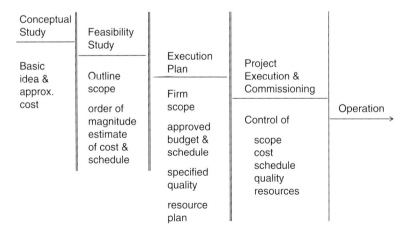

Figure I.A.3

_____

3 The RIBA Plan of Work 2013 comprises eight stages 0–7, detailing tasks and outputs required.

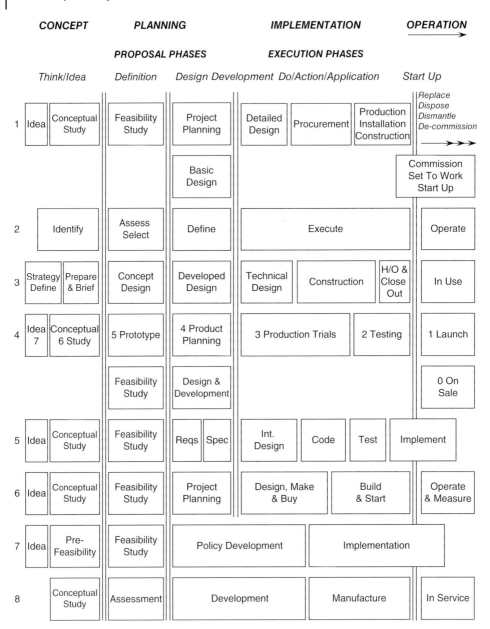

Figure I.A.4

2.2.1
Figure I.A.4 shows the different terminology and phase definitions used in different business environments:

Line 1 is a generic model for technological industries – process and power.
Line 2 is the owner/client perspective of line 1.
Line 3 is an architect-driven building project.
Line 4 represents product development.

Line 5 represents information technology.
Line 6 represents manaufacturing.
Line 7 represents The Civil Service 'Policy' project life cycle.
Line 8 represents the Ministry of Defence smart acquisition process.

2.3   Purchasing and Contracting Phases

There are three positions in the development of the phases where the owner may contract with someone else to perform the work in subsequent phases:

a. At the end of the conceptual study and start of the feasibility study
b. Prior to the development of the basic design and planning stage
c. Prior to the execution phases of the project

   The purchasing options available to the owner mean that the stages in Figures I.B.3 or I.B.4 involved in the contracting process have to be integrated into phases shown in Figure I.A.4.

3   Project Patterns

There are important patterns that depend on the phases of the project and give a clearer understanding of the way a project develops.

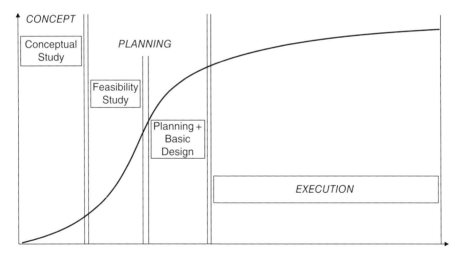

Figure I.A.5

3.1   Cost Impact of Decisions

During the feasibility study, alternative types of projects are being examined, and by the time the final study is accepted by management, the cost of the work is known to within a reasonable margin. Assuming the basic concept does not change, it is extremely difficult and often impossible to make more than a 15 per cent saving. In other words, 85 per cent of the cost impact has been determined during the front end phases (See Figure I.A.5).

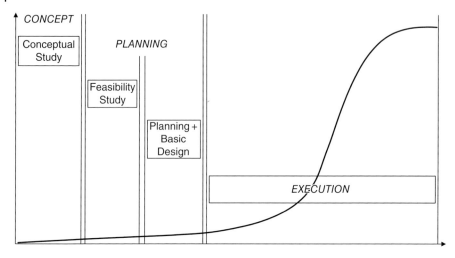

Figure I.A.6

3.2 Commitments

The financial commitment in the early stages of the job is very small compared with the costs once production work commences. It is much cheaper to totally change the approach to the project during the feasibility study phase when all that is involved is a new report than to make a change later when major equipment has been bought and work has started on site. This illustrates why it is important to have the best brains available during the early phases of the project (See Figure I.A.6).

4 Reasons for Projects

There are five reasons for doing projects:

a. Return on investment – only known after the facility is operating.
b. Achieving strategic objectives – both the public and private sectors.
c. Complying with legislation – safety, environmental, financial, and so on.
d. Political and social reasons or critical needs – question their validity.
e. 'Ego/Vanity' projects – particularly dubious internal projects wanted by a senior manager. If you can spot them, avoid them.

5 Project Needs

Projects cannot accomplish anything on their own. To survive they have needs that must be met. Namely:

a. They need clear objectives and complete definition.[4]

---

4 Software and business change projects need clarity of what is to be achieved, not necessarily their definition.

b. They need reliable finance.
c. They need political stability and certainty.
d. They need the shortest execution programme.
e. They need competent, capable, and experienced project managers.

Section B   Project Management Characteristics

At one stage during my time at Cranfield, I thought that it would be a good idea to research the definition of project management. I decided it would not be a difficult task since books on the subject would either provide a definition in their introduction or in chapter 1. Consequently, I looked up all the project management books in the Cranfield library and, to my amazement, they were all different!

For a brief time, as chairman of the Association of Project Management (APM) education and training group, I was involved in the development of the original APM body of knowledge, an excellent document that defines *what* the various subjects are that a project manager needs to know.[5] Theirs was another definition. On a quick glance at the U.S. Project Management Institute, I discovered that they define *how* project management subjects should be performed, with yet another definition. Too many definitions are complex (trying to cover every aspect of project management) and mix up projects and project management. This was when I decided that I needed to provide a definitive definition and, consequently, modified something that I came across in the paperwork in my office.[6]

Project Management is the multidisciplinary process of achieving a satisfactory end result.

The 'multidisciplinary' part (people working together as a team), creates complex relationships and a matrix organizational structure. It is what distinguishes project management from the individual functional disciplines. It is a work *process*. It is not a bunch of tools and techniques. The finite end result is the project, is always unique, and can be anything. Finally, successful project management does not have to produce the best; it just has to create something that is good enough, namely, satisfactory and on time and to budget. The purpose of this book is to help project managers achieve the necessary satisfactory end results.

Project management turns bright ideas into reality and is the means to achieving the end result and not the end in itself. Commitment to the project management concept is vital for the success of the project. Making things up as you go along is a route to financial disaster. Project management is the essential discipline that turns senior management's concepts, visions, goals, and strategies into practice. In June 2000 a survey in *Fortune* magazine showed that the single commonest reason for the failure of chief executives was their failure to implement their plans.

The challenge for the project manager is to manage complexity, ambiguity, uncertainty (risk), and urgency. In order to achieve success, a major effort must be mounted by all involved parties in the front-end planning of the work. Unless studies are carried out thoroughly and unless the planning is comprehensive and competent and unless

---

5 The CIOB has a *Code of Practice for Project Management for Construction and Development*, setting out everything that a multi-institute task force, with representatives from RICS, RIBA, ICE, APM, and CIC, has determined should be performed on a project. Published by Wiley-Blackwell, 2014. ISBN: 978-1-118-37808-3.
6 I have adopted a 'back to basics approach' rather than getting into detailed definitions which are covered by various ISO Standards, e.g.: ISO 21500:2012 Guidance on Project Management. There are similar standards for programme management and portfolio management.

the organization conforms to the standard requirements of the project management process, it is extremely unlikely that the project will be a success.

Language is the first barrier that deserves mention in the project management business. The project world uses the same words but applies them quite differently (see Section A, paragraph 2.2.1; Figure I.A.4). For example, I use project launch for the start of a project since the term *start up* is used in the process industries for the stage when their facility is set to work. However, in the product development business, *product launch* is at the end of the project when the product is being introduced into the marketplace. Similarly, I used *implementation* for the stage when the bulk of the work was carried out, but the information technology world uses implementation for the setting to work stage. Consequently, *execution* has been used for the carrying-out/doing stage. I avoid the use of the word *development* since it tends to bridge the last stage at which the project can be abandoned. (See paragraph 3.2.1 in this part and Figure I.B.5).

The second barrier in the project business is a cultural barrier. This cultural barrier is not just that between the French and the English, but it is between the various project management industries, where there is a reluctance to borrow good methods from each other. There is also a barrier between companies in the same business environment. One company will be design-dominated, another project management-focused, and another will be experienced in pharmaceuticals or in offshore work. Then there will be the companies that are the favourites of a particular client. If that lot is not bad enough, there is the cultural barrier within companies – the different mindset between the front end creative people and the practical back-end applicators. It is now recognised that project management is an attitude of mind, and this is what makes it more of an art than a science.

There is a potential third barrier that requires skilled project management. In the project management process, there are interfaces where conflict can occur quite naturally, namely, between:

- Client and contractor
- The main functional departments of design engineering, procurement, and construction
- The individual design groups
- The line functions and the task force

There are two components to achieving the successful end result. Firstly, the hard subjects, the 'hardware' of project management:

a. Strategy, contract and organizational
b. Financial analysis
c. Planning and scheduling
d. Control techniques
e. The four techniques that are the science of, and special to, project management:
    i. *Product and work breakdown structures* (P&WBS) should, on the whole, always be done manually as a team process.

ii. *Critical path method using network analysis.* The simple time analysis can often be done equally well manually, owing to the intricacies of the multitude of software packages. However, once the network is over a certain size it is safer to use a computer owing to people's ability to add. However, a computer will always be needed for serious project management if the real benefits of iii) are to be achieved.

iii. *Resource analysis and allocation.* Each type of resource (people, materials, money and so on) requires a different 'calendar' (working hours, shift patterns, shipping times, holidays and so on). Consequently, a computer is required for the complex analysis involved.

iv. *Progress measurement using earned value and 'S' curves.* This is the sophisticated part of project management and is avoided by many people. Further, owing to the effort required to implement the process it can often be compromised by simplification. (See Part V Section L).

Secondly, the soft skills, the 'software' of project management:

a. Teambuilding
b. Leadership
c. Communication skills
d. Presentation skills
e. Motivation
f. Influencing
g. Negotiating

These elements have all been borrowed from the toolbox of general management. As we can see, there is a conflict. Is project management an art or a science? The skill of the project manager is to decide where to put the emphasis – onto the hardware or the software – and how to integrate the two. I have absolutely no doubt that the more one is involved in project management, the more one is surprised at the power of these soft skills. Without the software, the hardware will achieve little. However, without the foundations of some of the hardware of project management, the efforts of the software is dissipated, and failure will result.

The clever part of project management is that, as well as modelling the project (the phases), it uses models of the various processes. This enables the project manager to evaluate different options before having to commit to specific actions with their associated costs.

1   Models

1.1

The overall project management process is broken down into the discrete project management fundamentals and modelled, using whatever management tool is appropriate.

Process	Model
Defining the scope and scope of work	Product & work breakdown structures (P&WBS)
Identifying the risks	Risk breakdown structure (RBS)
Fitness for purpose	Specifications
Risk and responsibility allocation	Agreement or brief
Forecast of costs	Estimating
Effect of changing variables	Risk and sensitivity analysis
Time value of money	Net present value (NPV)
Execution plan	Critical path network
Timing of activities	Bar charts
Project team	Organization structure (OBS)
Leadership	The project manager

1.2

As well as these individual models, there is an overall project management model (see Figure I.B.1)[7]:

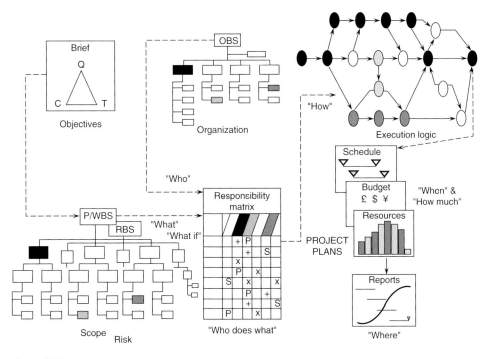

Figure I.B.1

---

7 The original version of this model was given the title 'The Project Model,' by Stone & Webster. I have developed it further and correctly described it as 'The Project Management Model'.

1.2.1

As can be seen, the model is composed primarily of the hardware techniques; all of the software being in the organization breakdown structure, with the communication links shown by the dotted lines.

1.3

In the model the *Brief,* or contract, defines the requirements and objectives of the project in terms of cost, time, and quality and determines which aspect will dominate the decision process.

1.4

The *What*, the scope of work, is defined by means of the product and work breakdown structures in order to identify manageable packages of work.

1.5

The *What if* identifies the risks using a risk breakdown structure developed using the product and work breakdown structures.

1.6

Who will lead the project, and who will form the team? The organization breakdown structure, is achieved by matching the requirements of the project and the abilities of the individuals.

1.7

Who does what transfers ownership and responsibility to the team (for example: full time – X, part time – P, support function – S) and communicates this through the responsibility matrix.

1.8

How the work should be performed is created through a team consensus for the execution plan and the relationship between the work elements.

1.9

When & how much is determined by the control documents that will provide the data (schedule, budget, and resources) from which trends and deviations from the plan can be identified and reported.

2   Characteristics

The project management *process* has certain characteristics, which differ from conventional management systems and brings with it certain advantages.

2.1 Project Management Objectives

The project management process takes the owner's or client's business objectives and translates them into specific objectives (Figure I.B.2) for managing the project (scope):

- By a specified time
- Within a specified budget
- To meet a specified standard of performance, which must include safety, other aspects being quality, value, and benefits

Figure I.B.2

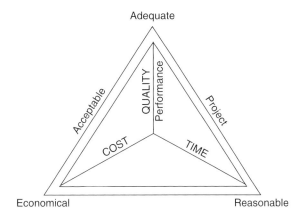

Since the prime objective is to complete the scope safely within the constraints of cost, time, and quality, some people put safety and scope at the centre of the triangle.

The natural instinct of owners is to ask for the lowest cost, the client project manager wants the shortest schedule, and the users want the highest quality.

All three of these extremes are not possible all together. The client should be asked to put an 'x' within the objectives triangle to show where the balance is. Do not accept an 'x' in the centre.

An insufficient budget or running out of money or missing schedule targets means that the scope of the project is compromised or cut. This results in a project that fails to achieve its objectives, and the consequence is dissatisfied users.

2.2 One Leader with Responsibility and Authority

A seminal requirement for any endeavour is that there is *one* person in charge; a single point of contact. Thus, senior management delegates the responsibility for managing the project to a project manager. The project manager is responsible for client relations and represents the client within their organization, and represents their organization to the client, as well.

In spite of this, in many organizations, senior management often says: "But we will make the decisions." In these circumstances, you have to ask who is managing the project: you the project manager or senior management. Consequently, you need to manage upwards – "May we have a decision by day x? Otherwise it will cause a delay to the project."

2.3 Multidisciplinary Teamwork

The primary characteristic that distinguishes project management from 'ordinary' management is that it is multidisciplinary. The difficulty is how to get these different disciplines to work together as a team. If it is achieved, there are two significant results:

a. Teamwork.
   The personnel on the project are more motivated and communication is greatly improved.
b. Synergy.
   The use of people with different skills, expertise, and experience to solve complex problems results in greater efficiency and innovation than ordinary groups could achieve.

2.4 Matrix Organization

Because of the temporary nature of projects, a matrix organization is necessary to:

a. Reassign personnel
b. Carry out long-term personnel planning
c. Audit the quality of the work

Organizations vary from the functionally organized with a project coordinator (a weak matrix) to the task force with a project manager (a strong matrix). See Section D, subsection 2.

2.5 Control

Project management achieves control of a project by ensuring that meticulous attention is paid to every aspect of the job.

3 Key Management Decisions and Phases

There are a number of key management decisions relative to the project phases.

3.1

Firstly, the owner may need to purchase additional expertise or resources at different stages (the phase breaks between feasibility, planning, and execution – Figure I.A.3) to perform portions of the work, as mentioned in Section A paragraph 2.3.

3.1.1
The first purchasing/contracting option is to negotiate with a contractor. This will have the minimun impact on the project duration (See Figure I.B.3).

3.1.2
The second purchasing/contracting option is to invite competitive tenders for the performance of the work.

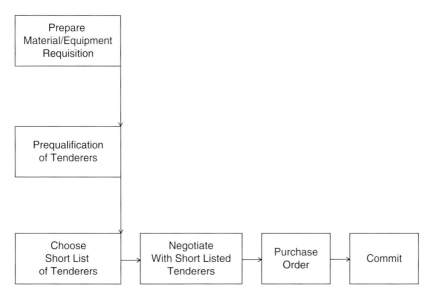

Figure I.B.3

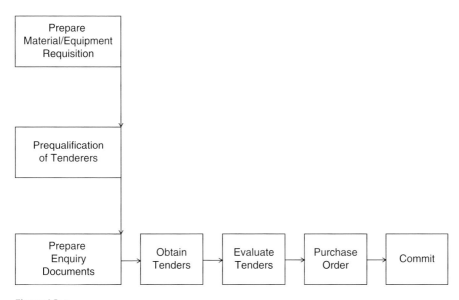

Figure I.B.4

3.1.3

The client may wish to retain complete control over the work by approving the deliverable outputs from each of the stages in Figure I.B.4. Consequently, an *owner approval* stage must be scheduled at the end of each stage. This option will significantly extend the project duration.

3.1.4

The owner/client may use the same contractor for every phase/stage or, alternatively, may use one contractor for the basic design and planning phase and then be tempted to invite competitive tenders for the execution phases. It is crucial for the contractor to maintain good relationships if they are to survive the transition from one phase to the next.

3.2

Project management (as well as deciding the contracting strategy in the early phases), must satisfy the criteria to move from one phase to the next phase, namely:

a. Is the project still appropriate to the company business plan?
b. Is the financial model still viable?
c. Will the project work technically?

3.2.1

The first two sets of vertical lines in Figure I.B.5 indicate where there is a natural break between the phases and an opportunity to stop the project. At the third set of vertical lines, the break is less natural, and the project can drift into the execution stage without proper evaluation. If the project cannot pass these 'decision gates' the project should be killed off. This is one of the most important decisions a project manager has to make and one of the most difficult to implement.

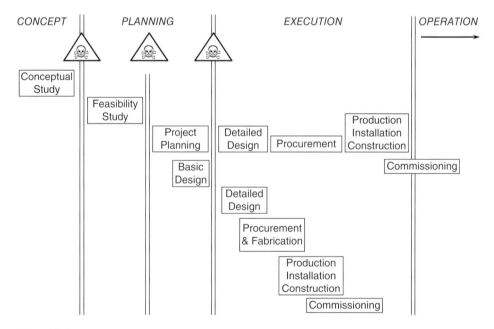

Figure I.B.5

3.2.2

Product development introduces more and more formal senior management gate reviews before the start of each numbered phase shown in Figure I.A.4.

3.2.3

For a client or sponsor, the gate review process starts at the identification of a business opportunity, and the most important gate is the one before any contracting arrangements are implemented. The last gate is a review of the project's success and the lessons learnt.

3.3

Thirdly, at the end of the planning phase, project management must decide how quickly to move into the execution phases and how much overlap of the phases there should be.

3.3.1

Starting the next phase before the previous phase is complete, *fast tracking*, means that rework will be required. The client may perceive this as contractor inefficiency, and yet the client is the person to benefit from an earlier completion date.

3.4

The last key decision is when to start construction. Just because the programme says one should be starting does not mean that one should start if you are not ready.

3.4.1

Some simple rules of thumb:[8]

a. Construction can't start until engineering has reached 30 per cent.
b. Construction can't achieve more than 30 per cent if engineering has not reached 90 per cent.
c. Between 10 per cent and 90 per cent complete construction can achieve 1 per cent progress per week. Less than this means that something is wrong. More than this means the key people in the project's management are deluding themselves. Ask what special plan or short cuts are being implemented, for example, pre-ordering of long lead items, extended working, using dedicated shipping, and so on.

3.4.2

Once construction begins, the construction people start demanding information, often in a different order to that which the home office is working. This must be controlled. Construction must be reminded of how they said they were going to work when their representative was involved in the design process (see Part IV, Section Q Installation and Construction, paragraph 1.3). Otherwise chaos will result.

---

8 Vernon T Evenson, Project Manager.

4 Project Management Patterns

4.1 Number of people involved

Projects have a definite start date before which the staff level is zero. The project manager has to find the resources required to get the project going. Consequently, projects start slowly. Conveniently, the number of people involved in the early phases, that is the study and planning phases, is very small compared with those required during the execution phases, see Figure I.B.6.

Once construction or production starts, the number of people rises very rapidly. The personnel build up to a peak and then fall back to end again at zero at project completion.

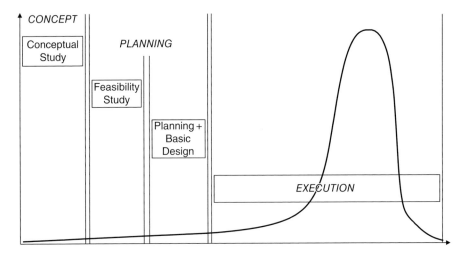

Figure I.B.6

4.2 Increase in costs for one week's delay or cost of accelerating project by one week

If the project is delayed for any reason, the cost of a delay during the early phases is relatively less expensive because fewer people are involved and few, if any, commitments have been made. If the job is delayed when the workforce is at its peak and a major part of the investment has been committed, the cost of a week's delay is very high, see Figure I.B.7. Thus, if a major design error is found, which involves rework, the cost impact can be serious. This highlights the importance of good design quality assurance.

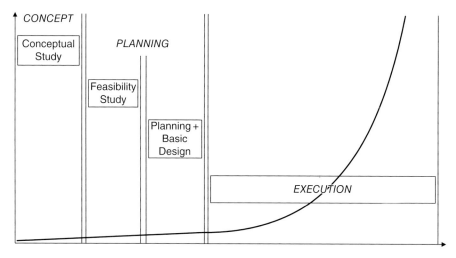

Figure I.B.7

Section C   Execution Planning Influences

As mentioned in the introduction, the concepts, principles, and processes of project management are the same for all projects. However, total flexibility is required in a project's execution.

Technological projects (the primary focus of this book) have a physical end product that is easier to describe and quantify and is easier to see being produced. Their success can also easily be measured. Information technology projects (see paragraph 4.8 below) are more difficult to describe, and the production process is not visible. Business change projects or programmes (see Part II, Section B) are, perhaps, the most difficult to visualise, and it requires considerable committment to realise their benefits.

This section identifies significant inbuilt cultural characteristics, as well as some imposed influences, on how a project is executed in different contexts. The language barrier mentioned in Section B is increased because a different language has to be used to make it acceptable/suitable in that environment. For example, the legal world is uncomfortable with project management but can cope with 'matter management', and risk analysis becomes 'due diligence'. In the medical world, the project manager is a 'consultant'. In the film world, the project manager is the 'director', and a breakdown structure becomes a 'story board'.

1   Project Characteristics, Size, and Complexity

1.1

Firstly, projects with multiple work fronts and access points: power, process plants and many civil and building projects, are ones where the sequencing of work has some flexibility.

1.1.1

Pipelines, railways, and road construction can also have multiple access points although less so. However, there is a very distinct sequencing to the application of different stages of the technology. An undersea pipeline, on the other hand, is more of a single workfront project.

1.2

Projects with 'tower project' characteristics means that the technology forces the sequence of the workfront, as in the initial stages of tall buildings. A true tower project would be climbing a mountain.

1.2.1

The existence of a bottleneck or 'pinch point': environmental or infrastructure limitations that constrain the delivery of materials or the disposal of excavated waste material.

1.3

Mega projects require a more collaborative approach between the contracting parties that may be outside the strict terms of their contracts.

1.4

All projects can be divided into natural components that have different characteristics, namely: the core technology unit, the support elements, and the surrounding facilities. Another approach is to divide the project into its cultural units.

1.5

Runners, repeaters, and strangers. See Section A Project Characteristics and Phases.

2 Strategic Decisions

2.1

Do not try to manage a large project. Break the project into manageable chunks, using a product breakdown structure and manage a collection of smaller projects.

2.1.1
The number of sub-divisions (contracts) should be limited by the same principle as the number of people reporting to a manager (five to seven).

2.1.2
The type of contract (see contract choices paragraph 2.4), competitive tendering and the incentive mechanisms to be used all have a major influence on project execution.

2.2

Who is to be the performing entity for each sub-division? The choice of performing entity is influenced by the organization's maturity, their skills and resources, and any proprietary technology that they own. Choices for the performing entity are:

a. The owner
b. A consultant
c. A joint venture
d. A contractor or subcontractor
e. A vendor

2.3 Organizational Options

Section D, subsection 2 details the different organizational options for the project management role. In summary the different ways for performing the work are:

a. At one end of a spectrum, information flow can be provided by a handover file being passed in a relay through the organization (not an effective option). Consistent communication can be provided more effectively by a project expeditor (the beginning of elements of project management).
b. More sensibly, continuity is provided by one person who is in charge, either full time or part time. A project manager, responsible for several smaller projects, will

naturally be part time. A project coordinator will provide unity of control, and a project manager will provide unity of direction.

c. At the mega project end of the spectrum, unity of command will be provided by a project general manger where the responsibilities will be divided amongst a hierarchy of project managers.

2.4   Contract Choices

2.4.1

The type of contract for performing each sub-division of the project should depend on who is best able to manage and control the risks and who is best able to estimate and carry the risks. The choices are:

Contractor Managed Risk:

Guaranteed Maximum

Lump Sum[9]

Fixed Price

Firm Price

Shared Risk:

Target Cost

Bills of Quantity

Unit Rates

Remeasure

Day-works

Time & Materials

Client Risk:

Cost + Fixed Fee

Cost + percentage Fee

This list of contracts has been itemised, in descending order, of risk to the contractor. From a client perspective, the order in each risk category should be reversed. The analysis is based on the named payment mechanisms being taken as written. Every client will have their own quirky way of administering them so that they can end up being quite different from the name they have been given. On one occasion, a contractor asked for my opinion on a contract that ended up being the complete opposite of what the name implied because of the obligations and consequent liabilities written in the

---

9 The I Chem E use this term in the same way that I use fixed price; namely, the price is fixed and does not change for a defined scope.

contract document. It cannot be over-emphasized that the true risk category is not revealed by the designated payment labels but is disclosed by the words in the terms of the contract. So read the words.

2.4.2

The choice of contract type involves balancing the degree of client involvement (reimbursable contracts) against the loss of client control (fixed-price contracts).[10]

The characteristics of the different contract types are summarized in Figure I.C.1. The arrows indicate that a particular characteristic is maximized in the direction of the arrow.

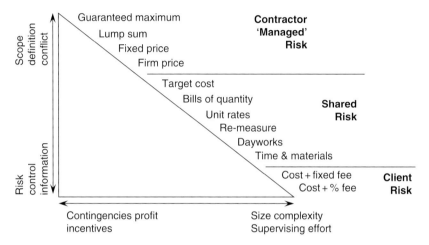

Figure I.C.1

2.4.3

'Tell me how someone is paid, and I will tell you how they will behave'. This is equally true of contracts. All contract forms and incentives have been developed to change people's behaviour. The two basic forms of contract are at the opposite ends of a behavioural spectrum. With a fixed price contract, the contractor will try to cut corners. Whereas with a reimbursable contract, they like to get involved in the detail and over-design in order to expand the man hours.

2.4.4

Traditionally contracting has been confrontational. The fixed-price contract says: "Leave us alone, and we will deliver an end product." The reimbursable contract says: "Tell us what you want us to do, and we will provide you with the necessary service." Today's emphasis is to build on this collaborative approach and work together for mutual benefit. Thus the choice of contract should also be based on inducing the right behaviour.

---

10 The introductory notes of the I Chem E, 'Model Form of Conditions of Contract for Process Plants', provides an excellent comparison between the two basic types of contract at the extremes of this spectrum.

2.4.5

If we take risk as the financial impact to the contractor, then a guaranteed maximum contract is the highest risk since it is in effect a fixed-price contract, but the contractor only keeps 50 per cent of any saving, whereas with a true fixed-price contract, the contractor keeps 100 per cent of any saving. Lump sum, as used by some people, means that you don't get paid your 'fixed' price until the project is completed, as opposed to a conventional fixed price, which can involve progress payments. A firm price removes the significant risk of inflation and as such is the lowest risk of this 'fixed' price, contractor risk category.

2.4.6

In the shared risk category, the client basically says: "We haven't decided on the scope yet, so please provide some rates for the listed work activities and the material quantities. When it is complete we will re-measure the work involved and reimburse you according to the schedule of rates."

2.4.7

The target cost contract is perhaps the only contract type that attempts, if structured correctly, to change behaviour. It can be seen as a lower risk since any overrun is shared with the client. However, any underrun is also shared, and consequently, has a higher financial impact than the other shared-risk contracts in which the contractor gets paid for all the work that they perform. There is not much to choose between the rest of the shared-risk category, apart from a bill of quantities. A bill of quantities is likely to include some additional risks in the words of the document, in addition to the pricing risk. Of course, if the contractor is asked to take any additional risk involving the quantities or scope, then this contract would move into the contractor-risk category. Day works and time and materials are basically the same. They should only be used by a client for unforeseen, unscheduled, and un-priced work. As a consequence, they can be quite lucrative for the contractor; however, the quantity of work should be small.

2.4.8

In the client-risk category, the contractor is reimbursed their costs so that there is no pricing risk as in the other categories. No contractor should be allowed a cost plus percentage fee contract, but I still come across them now and then. If you have one, look after it!

2.4.9

There is also an option to start the project with one of the types of reimbursable contract and then at an appropriate stage, when the scope of work is adequately defined, ask the contractor to convert the costs into a fixed price.

3 The Historic Nature of an Industry

3.1

An example is UK government projects: officialdom likes to pass responsibility but retain control, with a tendency to select the lowest price.

3.2

The culture of the owner's organization. An example is a declared preference for fixed-price contracts but with projects usually executed on a reimbursable basis or clients with a preference for a particular contractor.

3.3

Allowing the architect to be an external professional advisor rather than integrating them into the design team.

4   The Characteristics of the Industry/Business Sector

4.1   Engineering Construction Projects:

4.1.1
These projects are characterised by a visible and physical end product involving heavy capital investment. There is a primary emphasis on safety and the physical environment. A range of multidisciplinary skills are required, using a wide range of materials.

4.1.2
Offshore projects are really process plants mounted on a large support structure. Apart from lifting one onto the other, their prime difference from an onshore plant is that everything has to be prefabricated or modularised due to the exorbitant costs of working offshore.

4.1.3
Construction is different. Although much of the work is similar from job to job, unlike home office design and supplier manufacturing, it does not have a specific workforce or a specific work location. One contractor never builds the same plant at the same site for the same client with the same labour force. For this reason, the concept of progressive, incremental improvement which may be achieved in manufacturing, is not feasible in the field. "When a Japanese car factory has a quality problem, they go solve it; when a British engineering construction project has a quality problem, they write a procedure."[11] At least the implementation of a proper procedure for each activity will lead to steady improvement.

4.2   The Civils and Building Industries:

4.2.1
A key feature is the historic separation of design and construction. The architect defines *what* is required but rarely talks to the builder who is responsible for *how* the work is constructed, thus compromising the outcome. Nevertheless, specific contracting approaches (for example: design and build) are now used in order to address this deficiency.

---

11 Modified quotation from a U.S. manager.

4.2.2

Another key feature is that there is no single point of responsibility. The quantity surveyor is cost focused, and the architect is design focused. Architects aim for the highest architectural standards even when it is inappropriate in corridors and service areas. Further, as professionals, they give advice and do not assume any liability or responsibility. The traditional contract conditions nominate 'the engineer' (a client employee!) as an arbiter. On the other hand, a project manager makes decisions (as in the new engineering contract) for the benefit of the project, balancing aesthetics against practical considerations and cost.

4.2.3

The nature of the work is primarily civil with other disciplines having a reduced input. The emphasis is usually on cost, with attention paid to quality due to the high cost of remedial work. By contrast a modern building, for example Terminal 5 at London's Heathrow airport, can be closer to a process plant with 70 per cent of the cost being mechanical and electrical services.

4.2.4

Comparison of Engineering and Building:

Engineering	Building
Plant 80 per cent	Plant 40 per cent
Measured performance	Perceived performance
Dominance of function	Dominance of architecture
Technical complexity	Aesthetic subjectivity
Setting to work problems	Finishing problems
Expert client	Inexpert client
Centralized management	Decentralised management
High management overhead	Low management overhead
Cost engineers	Quantity surveyors
Man hour control	Cost monitoring
System design	Scheme design
Integrated design	Disintegrated design
Detail design by contractors	Detail design by design team
Low design cost	High design cost

4.3   The Power Industry:

4.3.1

This is a mature industry. Consequently, there is a strong focus on the specification and detailed design. The technology is 'common art' with little development required. The consequence is that open competitive tendering for turnkey, fixed-price or lump-sum contracts is common. There is a heavy client involvement with independent consultants. Projects are usually financed by loans or aid agencies.

4.4   The Utilities:

4.4.1
Historically the pre-privatization water industry in the UK was dominated by a civil engineering culture but has subsequently accepted that it is a process industry. They started with a lack of project management skills and have subsequently developed their project management philosophy from a variety of Industry practices.

4.5   Aerospace:

4.5.1
This industry is characterised by progressive development up to a prototype stage on a reimbursable basis using one or two contractors. These contractors then tender fixed prices for the execution phase.

4.5.2
Aerospace tends to be a series of repetitive projects managed as a programme of work. There is, however, the problem of timescale. They are too long for consistent personal commitment to get to the end, and project managers will change.

4.6   Government Projects:

4.6.1
Governments tend to believe that they are purchasing an end product rather than project managing the development of a project. Projects can lack one point of control, and the project management is weakened by the power of a contracts department. There is a tendency to top up their level of ignorance as personnel are moved onto other assignments and new personnel are brought in who have no project management expertise.

4.6.2
Try to avoid what can be termed political projects. The rules of engagement tend to change, and the project manager gets stuck with the problems.

4.7   Armed Forces Projects:

4.7.1
These are usually programmes of incremental capability acquisition as described in Part II, Section A. They are subject to the external influences of politics and have too many internal organizational elements.

4.7.2
A major problem is the advances in military technology that takes place during the long development stages of the project. For example: an armour-piercing projectile is developed. However, the opposition develop better armour or introduce sloping armour, so further development or new concepts are required.

4.7.3

Techniques need to be developed whereby key technology elements can be extracted in modules and changes retrofitted.

4.7.4

The major flaw in defence projects (as with all government work) is that there is too much concentration on the procurement process and not enough on project management.

4.8 Information Technology Projects:

4.8.1

The industry is still relatively young. The characteristic that distinguishes it from other industries is that the hardware involved, the design techniques, and the software languages change very quickly. However, the developments in the software have not kept pace with the potential of hardware. This encourages people to add extra functions to the requirements. It is also poorly understood by management. The project managers are rarely trained, and the creative nature of the project personnel produces organizational behavioural problems. The IT discipline seems unable to accept that conventional project management has anything to contribute to the management of their projects. The discipline has a habit of reinventing 'new' management processes. The biggest problems are not technical but human problems.

4.8.2

There are control problems due to the ongoing nature of the design process and the lack of a visible product. Control is exercised through man hours spent, lines of code written, or modules completed. A problem is that there is a tendency for everything to be 99 per cent complete. Finally, as for many projects, the documentation – operating instructions and so on – tend to be deficient.

4.8.3

The phases are very similar to any engineering project, but software development has some differences:

a. As in many projects, the customer may not know what they want in the initial phase of *analysis,* which produces a *requirements specification.* Further, the customer's ideas may change when they see what is proposed and also during development. Again, as for all projects, their business needs may also change. Consequently, there is a formal process of *validation,* which checks that the requirements specification is correct.

b. During *design,* there is a formal process of *verification* to check whether the design meets the requirements specification. Prototypes of the proposed system may also be produced. The design stage must also include a specification of the hardware require-ments.

c. *Implementation* of the actual coding of the programs and the like is probably about one quarter of the total project effort. Similarly *integration and testing* requires at least as much effort as implementation.

d. A *maintenance* process is necessary since it is virtually impossible to test for every eventuality in a major system. Software must also be designed with appropriate consideration for possible future enhancements. Control of both documentation and change is vitally important.

e. There is no procurement (or none to speak of, though part of the system might be brought in). There is no construction.

4.8.4

Software development has been much concerned with the possibility of software re-use. Thus there is a need to develop standards so that the input/output interfaces and performance of the software component are well-specified. This is like buying a pump; one would buy it based on its performance specification without knowing much about how it does it.

4.8.5

The information technology world is beginning to relearn that large projects cannot be managed. Contrary to conventional wisdom (for physical projects), it is a mistake to try and define the total scope for a large software project. It is clarity of the objectives, of what one is trying to achieve, that is critical. Decide on the division of work by breaking the project into its natural smaller packages. Priortise the component parts that provide the best value for the users and hence the business, and release incremental functionality early. Roll out the various elements slowly in order to find problems before releasing the next component. 'US government statistics show that the majority of [IT] projects that run for longer than a year with no intermediate deliveries never deliver, no matter how convincing their business cases are'.[12]

4.8.6

Start with an extensive feasibility study and a pilot project that is thoroughly tested by a small group of experienced users, and then grow the project as it is released to a wider population of users. This approach was demonstrated and validated with the computerization of pay as you earn (COP) project – 1977 to 1983,[13] The first large government IT project completed on time and under budget.

4.9   Change Projects

4.9.1

The story of a survivor from the Piper Alpha North Sea Platform disaster provides the key to any project involving change. When the survivor was asked, "Why did you jump into the burning sea when you were probably jumping to your death?" His reply was, "There was a chance I would survive, which was better than certain death by staying on the platform."

---

12 Top seven agile behaviours that result in success' by Brian Wernham published in 'The Agile Business' by Raconteur Media 14th October 2014 in The Times and posted in APM Resources as 'How to manage agile' 6th January 2015.
13 *Project Managed* by Steve Matheson. His is a name that should be on the role of honour of the project management world.

4.9.2

The business parallel is to recognise that if the company carries on as it is, it will slowly die, but if it changes, it might survive. Ask the questions: "Where are we and where do we want to be?" The cost of staying where we are is going to be greater than the cost of the process of change. Any project involving a change in cultural values must also involve changing people's behaviour by changing the company's procedures. Both have to be altered to achieve change that will last.

4.9.3

The absolute commitment of the chief executive is essential. They must also stay committed throughout the project. If they are not involved, don't start the project. I was appointed project manager of a productivity improvement programme; it was launched by the chief executive and because of their presence at meetings, even the most negative of the department managers (one in particular) also attended. However, at about meeting four, the chief executive excused themselves because they had to meet a client. The message received by the reluctant attendees was: "Oh, so there is something more important than this time-consuming meeting." The result was that the sales director also had a client to meet at the next meeting. From that point on the whole programme fell apart.

4.10   Manufacturing:

4.10.1

Manufacturing projects are different in that they perform repeat projects for the same client in a constant location with the same labour force.

4.10.2

The key to a manufacturing plant is to identify the constraints (or weakest link) in the chain of activities (the process) that delivers a finished product outside the factory gate. The weakest link (bottlenecks) in the chain of activities (the machine with the longest component manufacturing time) determines the time taken to complete the assembly of the finished product. The focus is on controlling the level of inventory in front of machines (in particular, the bottlenecks) to the throughput (the finished products that have been sold, resulting from market demand).[14]

4.10.3

In the car component industry, the good news is that you have been awarded the contract. The bad news is that in year two we will expect a 10 per cent reduction and in year three a further 10 per cent reduction. We may then have to invite competitive tenders in case you have become too complacent or because of developments in the marketplace. This is the approach that anybody outsourcing services, such as providing canteen facilities or computer services, should adopt.

---

14 For a thorough understanding of these issues read *The Goal* by Eliyahu M. Goldratt and Jeff Cox, published by Gower.

4.11   Research and Development:[15]

4.11.1
It is incorrect to group research and development together. In a development project, a deliverable has been determined (see aerospace above), and thus it conforms to the norms of project management. Research projects, on the other hand, contravene most of the essential principles of project management.

4.11.2
In principle, a research project manager is responsible for supporting creative thinking in small subject-oriented units and making sure that the thinking results in some kind of concrete output. Further, this output should preferably be on time and to budget. However, if there are fixed goals with certainty of the results, then it is not research. Additionally, if there are no failures, it can be argued that the researchers are not doing their job.

4.11.3
Research is more project leadership than management; direction comes from the work itself rather than from a manager. It is to a large degree about influencing and persuading partners.The research leader's task is to present a unifying vision and nurture a project environment where an assembly of individuals can be turned into a committed and effective team. It should feel responsible for, not only their own individual contributions, but for the collective team output. Researchers have a desire for a large degree of autonomy in their work and democracy in decision-making, Hence, a high degree of delegation and attention to interface management is required. A democratic-authoritarian management style is needed. That is, *consensus with qualification*, see Part VI Section B, paragraph 1.2.

4.11.4
Listed below are some of the paradoxes of research project management:

a. There is a need for a risk-taking approach to be innovative. By contrast there is the need to reduce risks abd to ensure the delivery of the desired result on time and to budget.
b. There is task and process uncertainty. There is unpredictability of the research outcome. New research opportunities arise during the course of the project, as against the need for predictability of project output.
c. The quality of any output may improve if deviations from the plan are allowed. Continuous adaption and adjustment is required. Flexibility is necessary in order that the project goals can adjust to future changes in the project. That is, change is a good thing!
d. Focus on getting the right things done, not so much on controlling how and when they are done. Effectiveness is generally more important than efficiency in research projects; the result is more important than the process.
e. Rather than setting one common goal, the first phase should be about juggling several versions of the project in the air at one time.

---

15 This sub-section is largely taken from a paper 'Project Management theory and the management of research projects' by Erik Ernø-kjølhede, Copenhagen Business School, January 2000. ISBN 87-90403-70-3.

f. The phases should not be considered as a deterministic linear process where each phase succeeds the other. They consist of a number of fundamental project tasks that overlap and gradually take turns in dominating during the life of the project. The conceptual phase will continue to influence the project but with diminishing intensity.

g. Participants are more likely to have powerful hidden agendas: co-operating in a project but in competition with each other for acknowledgement of their contribution. They have a need for recognition of their work and to be published.

h. There is a lack of management information and a difficulty of interpreting management information. There is uncertainty of the end product and the process versus the need to act as if there is certainty and making management decisions continuously. Realistic planning is not possible in view of the high level of uncertainty.

4.12   Theatre and Film:

4.12.1

A theatre production is a series of repeat projects that continue for as long as they produce the business benefits for the client (the producer). Each repeat project will be slightly different, depending on the users' (audience) reactions. The business-oriented client tries to control the creative project manager (the director) who often has little interest in cost issues but works to a 'drop dead' (see 5.2 below) opening date. There is almost always an extensive feasibility and protyping stage (rehearsals and provincial tours before a major city launch). It is, in effect, a 'tower project'; it has to start at the begining and continues in sequence (as defined by the specification – the script) until the end is reached. Theatre is led by the creative team (actors) who use the site (stage) first and then the technical team come in to work around the actors, and adjust the scenery in order to complete the scope. Once the execution phase starts, the project manager has no more involvement and the team (cast) is self-managing. A theatre production is a project being produced live – the end product is transitory and, on its own, leaves no permanent physical mark.

4.12.2

By contrast there is a permanent end product for film projects. They are in effect pipeline projects with multiple access points; that is, they can be done in any order. Films sometimes leave out the feasibility stage (rehearsals) and take multiple shots at getting the execution stage right, which is the reason for cost overruns. However, if the quality is right, the business returns swamp the cost overruns. Films set up and do all the technical work first, and then the creative team (actors) come onto a finalized project location (set).

4.12.3

A film will perform a product breakdown geographically with a location focus. Even if the action takes place over hundreds of years, all outdoor sequences, indoor sequences, or all the scenes involving certain features, settings, or actors are grouped together for cost efficency. Film projects also involve large numbers of subcontractors and thus need much more formal and detailed planning. Filming individual scenes takes on the characteristics of a drop-dead date. Conversely, if they meet problems with the setting availability or resources outside their control, the creative director will change the specification.

The film *The Making of Gone With the Wind* demonstrates that *Gone With the Wind* was completed on time and to budget because it used a significant range of project management techniques and skills. Particularly noteworthy is its use of product and work breakdown structures, taken down to the task level, in the form of story boards.

4.13   The Medical World:

4.13.1
Projects in the medical world are unique in that they form part of the client's own body. For minors, the client and the key stakeholder are separated. The client's cost, time, and quality/performance triangle are totally skewed. In conventional project management, the client sets the objectives that must be achieved. In medical projects, the project manager (the consultant) decides on the objectives for the project, and the balance between time and quality and cost is now also considered. However, this is not done alone but in conjunction with their colleagues, the other project managers/consultants. For the client, the objectives are aspirational, and their whole emphasis is on quality with a secondary interest in time. In the National Health Service (NHS), the client ignores cost. For private work, cost will be a major consideration but will hopefully be covered by a project insurance policy.

4.13.2
The feasibility and planning phases of a medical project, for example a surgical operation, are dominated by intermittant tasks (pre-assessment, tests, and so on) in a multi-project environment. None of these early task performers understand the whole picture (project plan). All the tasks interact with each other, and the interuption takes precedence. In order to complete the interupted task, there is a reassessment and, consequently, the overall duration of the work takes longer.

4.13.3
Somewhere in the organization, there is a programme manger using a multi-project approach juggling availability and allocation of resources. The result is that the client/patient may never see the same team member more than once, performing the same task. Thus the client/patient is totally dependent on a functioning system rather than a single point of contact in charge.

4.13.4
The project (patient) is passed in a relay with a handover file from one specialist technologist to another, with little or no prior knowledge of the project. The handover file grows in size as the interactions accumulate and the project moves from one stage of the process to the next. There is an enormous emphasis on interface management to ensure that the correct project/patient is being addressed.

4.13.5
The execution stages are very similar, if not identical, to the model describing the conventional process in Section B. The main difference is in the people, and in particular, the

project management function. There is a manager of projects (the senior consultant), but they are also the head of a functional department, and they hold meetings with the project managers to agree the plans for the various projects.

4.13.6
Whilst I can make a good case for a project manager not being trained in the technology of the project that they will manage, this does not apply in the medical profession. The project manager/consultant has to have been trained through the route of technologist (the registrar). There may also be an assistant project manager (nurse practitioner) who will be permanently assigned to the project manager on all the projects that they are involved with. Whilst it is theoretically possible for an assistant project manager to eventually qualify as a consultant, this is unlikely. Whereas in the technical industries, this is more than likely. There is not a conventionally understood deputy project manager. The registrar is in effect a project engineer in training to be a project manager. Further, as on small technological projects the consultant project manager may also perform the technology/operation.

4.13.7
Another difference is in the project team for the execution phase. As a generalisation, a team is normally selected from whoever is available; however, for a surgical operation, the theatre team has been pre-selected and has worked together so that team building should not be necessary.

4.13.8
The main part of the project, the operation, is project managed as a *stranger* project with elements of a *repeat* project (see the first paragraphs of Section A). It is reassuring to see that "surgeons will use a pre-operation checklist:

Does everybody on the team know each other's name and role?
Has the surgeon briefed the team on the goals of the operation?
Do you have the right side of the body you're operating on?
Only then do you begin.[16]

4.13.9
The whole approach changes for the necessary postoperative tasks and ongoing support functions – *runner* projects. However, by good definition of the tasks and how they should be performed, many become standardized routines.

4.13.10
At the end of the project, the setting to work stage, the technological and medical projects are very similar. In both cases the project manager discusses with the client any short fall from the 100 per cent performance that the client wants, that can be or may have to be accepted. However, something that the medical world does well, compared

---

16 From an article in *The Times*, Saturday November 22, 2014. 'How a checklist saved a little girl's life.'

to technological projects, is the postproject completion stage when follow-up projects are discussed.

4.13.11

Accident and emergency projects are totally dependent on a good process (capable of a wide-ranging flexibility in capacity) with good systems and procedures. The first step is a pre-feasibility stage (triage) in order to determine priorities/urgency and filter the various technologies. Is it a pipework job, electrical, software, or is it a piece of machinery that doesn't work? At the feasibility stage, a preliminary assessment is made; is the project a runner, repeater, or stranger? Can it be repaired, do parts need to be substituted, or in the ultimate case, does the machinery need replacing – a major project.

5 Phases and Schedule

5.1

In the ideal project, all of the design is complete, and all of the materials are available before construction starts. However, once a schedule end date is imposed to shorten the schedule, work has to start on the next activity before the previous activity is complete (it has to be 'fast tracked'). The question is: 'how much overlap should there be?'

5.2

With the imposition of an immutable end date – a 'drop-dead' date, such as for the Olympics – the project will be set to work regardless of whether all the final details have been completed or are available. As a result, the project may be launched too early and compromise the planning process. Alternately, it imposes the major risk of urgency on the project process, which in turn is a major cause of project failure.

5.3

As already stated, running out of time means that the scope of the project is compromised or cut. This results in a project that fails to achieve its objectives and the consequence is dissatisfied users.

6 Execution Planning

Detailed execution planning issues are addressed in Part IV. However, some specific project influences, and options where decisions have to be made, are listed and summarised here for completeness.

Division of the Work by Phase	Considerations for the Performance of the Work
Phase:	How much overlap.
Design:	Done in-house. Subcontracted, use consultancy expertise (or an architect!).
	Using local knowledge/resources.
	Technology transfer involved.
	Legal requirements.
Purchasing:	Determine the level/amount of vendor data needed.
	Use proprietary information or catalogue information/data.
	Use vendor standard designed equipment or use own design.
	Buy bulk materials or get subcontractors to supply
	Identify where no expediting is carried out.
Expediting:	Identify where it is done by telephone.
	Identify where visits will be required.
Inspection:	Decide when goods, materials and equipment will be inspected.
	Upon delivery, at the suppliers'
	During fabrication.
Construction Management:	Is it to be cost or schedule dominated?
Construction:	Direct hire labour or subcontracted.
	Use of local resources/contractors
Commissioning, Start up,	Get the users involved early.
Setting to work:	Develop the 'start packs' during construction

7 Generic Influences on Project Execution

a. Competitive tendering and incentives
b. Legal requirements
c. Local knowledge and/or consultancy expertise
d. The need to use local resources/contractors
e. Designed equipment

Section D   The Project Management Role

The project management role is complex and can vary a great deal; however, the role of the project manager has fewer options. Both roles can be defined under the following groupings:

- Strategic and contractual
- Organizational and functions
- Responsibilities and orientation
- Competencies and leadership
- Abilities and skills

1   Strategic and Contractual

An owner has a number of management and contracting options for managing a project. They can use:

a. An owner's team to manage a main contractor, using any one of a number of contract variants
b. An owner's team to manage a number of subcontracts, again using similar or different contract arrangements
c. A consultant or architect (not recommended) to act on their behalf, instead of using their own management team
d. A contract for a project management services organization to supplement their own core management personnel

2   Organizational and Functions

Each of the above contracting arrangements can have different organizational options for the project management role.[17] Should the organization face the outside world saying, "This is how we manage our business through disciplines and functional groupings", or should it face its customers saying, "This is how we deliver end products through project management"? These two axes at right angles to each other have different agendas. The axes can then be rotated so that one role is more dominant than the other. Thus a matrix of mixed organizations is formed that can merge into one another as the project team grows and declines. The main structures are as follows:

---

17 The definitive and seminal work on the subject of 'matrix management' is a paper titled 'Organizational Alternatives for Project Management' by Robert Youker, of the Economic Development Institute, World Bank, Washington, DC. Presented at the 8th Annual Symposium, Project Management Institute, Montreal, Canada, 6 October 1976. Everything that has been written since then has 'borrowed' from this paper. It says almost everything that needs to be said on the subject, although I tried to add to it in a paper I wrote for the Norwegian Institute of Technology 'Nordnet '91' conference, Trondheim, Norway, 3–5 June 1991, titled: 'Project Organization Structures from Logic to Reality'.

2.1   Structure A – Functionally Managed:

The project is divided into segments and assigned to relevant functional areas and or groups within functional areas. The project is coordinated by functional and upper levels of management. However, basically there is no project management as such, though an expeditor may be appointed to *improve communication*.

2.2   Structure B – A Weak Matrix:

A project manager with limited authority is designated to coordinate the project across different functional areas and or groups. The functional managers retain responsibility and authority for their specific segments of the project. The project manager tries to *control* cost and time.

2.3   Structure C – A Balanced Matrix:

A project manager is assigned to oversee the project and shares the responsibility and authority for completing the project with the functional managers. Project and functional managers jointly direct many workflow segments and jointly approve many decisions.

   This is probably the most difficult arrangement (but very common in business) in which to exist.

2.4   Structure D – A Strong Matrix:

A project manager is assigned to oversee the project and has primary responsibility and authority to *direct* and complete the project. Functional managers assign personnel as needed to a co-located team and provide technical expertise. This is commonly referred to as a 'task force' but is still strictly a matrix. This structure is the recommended and preferred option for serious project management.

2.5   Structure E – A Task Force:

A project manager or project director has sole responsibility. They are put in *command* of a project team composed of a core group of personnel from several functional areas and or groups, assigned on a full-time basis. The functional managers have no formal involvement. This structure is usually used for mega projects, and the project manager or director will probably divide their responsibilities amongst a hierarchy of project managers with different levels of experience.

2.6   Sub-structures:

There are similar choices at the lower hierarchial level of project management. The project engineering function can be organized in a weak functional matrix with the project engineers coordinating across the design discipline groups. Alternately, the project engineers can be organized as a strong matrix in charge of a group of design disciplines for their individual parts of the project. In some cases a separate technical task force might be set up for a stand alone (a complex equipment package) part of the project.

2.6.1

We lost a competitive tender becaust the client perceived a competitor to have *better* project engineers, managing a strong matrix of co-located designers. Our project engineers were coordinating across the designers in their functional disciplines, in a weak matrix.

2.7

The project manager does not manage alone. There are at least four distinct functions to be performed in each of the organization options:

a. The user or originator of the project
b. The client/owner or sponsor of the project
c. The project manager acting in one of the contracting roles above
d. The project team

Problems occur if any one of these functions is missing or merged.

2.7.1

The role of the users is addressed in Section G Achieving Success, subsection 3.

2.7.2

The client/owner project manager is different in that they take a more strategic, benefits-and-outcome perspective. The role of the client/owner or sponsor of the project is addressed in Section F The Owner and Client.

2.7.3

The project manager is the leader of the team who, regardless of technology, will involve at least the following functions:

a. The manager of the technology – in our context the engineering manager
b. A manager of the commercial aspects – the procurement manager
c. A manager of part of the project manager's function – a project controls manager (or project office manager) responsible for estimating, costs, schedules, and gathering data for reporting
d. A manager of the site execution phase – the installation or construction manager, who eventually hands over leadership of the project on site to the commissioning manager
e. On an international project, with multiple sources of finance, there may be a need for a project accountant

2.8

The project management function can sometimes be expanded with additional roles, such as a deputy project manager, an assistant project manager, and a champion.

2.8.1

A deputy project manager is, in effect, a project manager in training and will stand in for the project manager in their absence. The deputy project manager should be given responsibility for a meaningful part of the project or a stand-alone sub-project to manage. They can also take over from the project manager towards the end of the project. In these circumstances the client may well agree to release the project manager since they will have seen the deputy perform over a long period.

2.8.2

An assistant project manager is basically an administrative assistant role and does not have delegated project management authority.

2.8.3

A champion is needed for internal projects and is someone who has no direct involvement in the project. However, they are of sufficient seniority and experience to advise you, the project manager, on how to manage the superior and difficult stake holders.

2.9

For joint associations/venture projects and projects with significant political stakeholders (internal projects and some public sector projects) the project manager should create a project board or steering committee to act as the client.

Make sure the most anti-political manager (or organization representative) is on the board in order to defuse their antagonism.

3   Responsibilities and Orientation

3.1

The responsibilities of a project manager can be summarized in the following reasons for having a project manager. They are to:

a. Centralize in one person, *who has no other duties*, all the responsibilities for a project. The cost of project failure can be huge. The cost of a full-time project manager is relatively little. So, why compromise the project management function? Nevertheless, a project manager may be responsible for two or three small projects – one starting, one established, and one finishing.
b. Set realistic goals for the project for all participating groups, considering the resources that each can bring to bear.
c. Make decisions on the project quickly enough to meet its needs and benefit it as a whole, not just a portion.
d. Provide a means of anticipating the problems during the course of the project.
e. Give one person the responsibility of quickly developing solutions to problems so that the project will stay on target for programme time and budgeted cost.
f. Consider the ethical view of the project and have sufficient knowledge of ethics and the law to be able to communicate the issues, setting an example through integrity and honesty and projecting the reputation of the business.

3.2

These responsibilities require the project manager to be orientated through 360 degrees, in different directions and different roles.

a. *Up and Down*. Reporting to senior management and managing the team.
b. *Internally and Externally*. Working within the organization and managing the project participants. Interfacing with the client, contractors, and suppliers and other stakeholders.
c. *Backwards and Forwards*. Controlling and measuring what has been completed and planning the next period's work.
d. *Sideways left and Sideways right*. Interacting with functional managers and observing how other projects might affect their project.
e. *Present and Future*. Acting as a spokesperson for the project and anticipating how the finished project will impact the organization or environment.
f. *Take a Short Term View but Little Long Term Perspective*. Project managers are really only interested in completing their own project. However, they will have to work with some of the same people again in their own organization and the client's. The difference with the client project manager is that they take a longer-term view. They are interested in satisfying the business case and delivering the benefits of the project.

4 Competencies and Leadership

I have had over twenty years as a practicing project manager and twenty-five years as a consultant and trainer. Having trained over 4,500 project managers over a wide range of businesses and industries; I have concluded that the following are the primary competencies required by a project manager:

1. Leadership
2. Interpersonal skills
3. Problem solving
4. Personal qualities.

4.1 Leadership:

a. Flexibility
b. Delegation
c. Resolving conflicts
d. Team building

4.2 Interpersonal Skills:

a. Communication
b. Persuasion
c. Negotiation
d. Influencing

4.3 Problem Solving:

a. Analysis
b. Judgement
c. Decisiveness
d. Creativity

4.4 Personal Qualities:

a. Integrity
b. Self-confidence
c. Tolerance of ambiguity
d. Political awareness
e. Helicopter perspective
f. Proactive working style
g. Determination

4.5 Leadership

Leadership is so important that it deserves to be highlighted more than once. See Part VI, Section B for the different leadership and motivation models as follows:

Tannenbaum and Schmidt	One-dimensional continuum model
Blanchard and Hersey	Two-dimensional situational leadership model
John Adair	Action-centred leadership model
MBWA	Management by wandering around.

5 Abilities and Skills

The abilities required of a project manager are many; however, the details of the following four are the dominant ones that distinguish the project manager from the functional manager.

1. Ability to persuade/leadership
2. Commercial business sense/financial
3. Ability to take helicopter view
4. Problem-spotting/solving ability

5.1 Ability to Persuade/Leadership:

a. Sponsor or client relations
    i. Think about an after-sales service
    ii. Keep the client off the project team's back
    iii. Resolve language and communication barriers

b. Company and management relations
    i. Market the project internally
    ii. Get resources
    iii. Get management support
    iv. Act as a change agent
c. Get people to work together
    i. Team building
    ii. Act as a coach.
d. Create a good job climate
    i. Be effective
    ii. Be enthusiastic and have fun
    iii. Protect the team from blame; share the credit

5.2 Commercial business sense:

a. Get a good deal for their company
    i. Change order control
    ii. Negotiate a favourable contact interpretation
    iii. Get value for money from contractors and suppliers.
b. Get a good deal for the client or sponsor
    i. Manage suppliers and subcontractors
    ii. Get good tender prices
    iii. Ensure that work is done for a good price
    iv. Watch back-charges, return surpluses
    v. Judge impact of changes on cost and schedule
    vi. Get value for money from the company

5.3 Ability to take helicopter view:

a. Don't get too involved in detail; delegate
b. Take action; be decisive
c. Monitor everything
    i. Create an early warning system
    ii. Look ahead for potential problems
d. Be objective
e. Be clear thinking
f. Be unemotive

5.4 Problem spotting and solving ability:

(See Part VI, Sections J and K)

a. Spot problems early
b. Discuss and test alternative solutions
c. Choose a solution
d. Implement the solution
e. Check that it works

5.5 Some of the skills required of the project manager:

a. Motivational and interpersonal skills to lead and drive a team to achieve difficult targets
b. An aptitude to resolve conflicts at organizational interfaces through negotiation and diplomatic skills
c. Good communication and presentation skills
d. Good letter-writing and report-writing capability
e. A proficiency at project appraisal and financial techniques
f. Knowledge of the systems and techniques required for effective project control
g. A capability at planning and managing resources, time schedules, and budgets
h. A thorough knowledge of contracts and the contracting process and how to deal with subcontractors
i. Knowledge of the procurement process and how to negotiate with suppliers
j. An understanding of quality and safety programmes

6 The Project Manager

6.1

The importance of the project manager cannot be overemphasized. They can be the reason for a client awarding a contract to their organization and making the difference between success and failure of a project.

6.2

My analogy for the role of the project manager is the conductor of an orchestra. They know musical theory (project management). They have played and even been skilled in one of the instruments (disciplines) and know something about the other instruments. They have studied the score (plan) and decided what emphasis (objective) to give each section (work package). They know when to bring in each group of instruments (functions) and have fun leading the orchestra (team). They create a successful end product, appreciated by their audience (users).

6.3

One could produce a list of the duties of a project manager, but then one would miss something. Consequently, the duties can best be summarized as doing everything and anything necessary to deliver the project scope, safely, on time, to budget, and to the appropriate quality standards, and to have a happy client.

6.4

With good relationships, the project manager's project management can flourish.

6.5

A final thought: project managers are judged on how they performed on their last project. Consequently, our careers depend on better project management, which starts with improvements to ourselves – the project manager. Keep your continuing professional development (CPD) records up to date.

Section E   The Manager of Projects

Most projects are required to report on a monthly basis. One's first thought is that this is a monthly report to the client. However, there is a requirement for a separate internal report focusing on the financial aspects and anything that exposes the company to risk.

The data that will be reviewed will usually be in a company standard format in as concise a format as possible. The manager thus gets used to seeing the same information in the same position on the report. This tends to highlight figures that are outside the norm and makes them more obvious.

1   Financial Situation

1.1   Obtain a clear summary of the project's overall cost status and financial position.

a. *Actual cost incurred to date with the current estimate.* Check that actual cost data is up to date. Some company systems can be slow in allocating costs to a specific project.
b. *Change order status.* The value of changes agreed and any that are in dispute with the client. What is happening with the disputed changes?
c. *Cost trends and variances, with explanations.* It is rare that there are no changes, so be suspicious if everything is stated as being on plan.
d. *Forecast cost to complete compared with plan.* This requires careful scrutiny since project managers are optimistic. They tend to show the forecast to complete as the original budget minus the actual costs. For example: is the forecast of productivity or monthly progress figures to complete the work consistent with the experience to date? If not, are they justified?

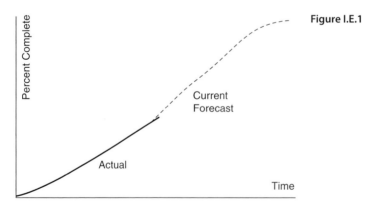

Figure I.E.1

Figure I.E.1 looks a reasonable forecast until you see the history in Figure I.E.2. This is taken from an actual project, and it was my intention to delete forecast f/c 1 since I thought that it might lack credibility. Forecast 1 demonstrates three things: (i) the optimism of project managers mentioned above; (ii) the importance of the launch phase (see Part IV Section A – Figure IV.A.1). and (iii) once the rate of progress is established, it is very difficult to change it. In this instance, the manager of projects

Figure I.E.2

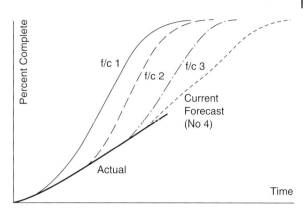

might have accepted f/c 1 on the basis that the project may not have started well, but now that the team has got its act together, progress would take off. However, the rate of progress has not changed, and at f/c 2, the manager of projects must ask, "Tell me what dramatic action are you going to take or what special team building are you going to implement in order to change the rate of progress?" Forecasts 3 and 4 should not be believed, and the forecast project completion can be determined by extrapolating the actual curve.

e. *Contingencies—their status and drawdown plan.* There needs to be a defined and documented basis for retaining contingencies and then releasing surplus contingencies to profit.

f. *Funded liabilities status* (liquidated damages, warranties, plant performance guarantees, and so on). These liabilities can be substantial and can make the difference between a significant profit or a significant loss. They should receive plenty of management attention, especially in the final stages of a project.

1.2   Bonds:

Bonds status, (see Part V Section P, Surety Bonds). Identify the type of bonds that are current. The amount of exposure must be stated together with their expiry dates. However, remember that in some countries, bonds never die.

1.3   Performance incentives status:

Consider the potential for a bonus for achieving targets, as well as the penalties that could be incurred by missing defined objectives. What are the plans for achieving each objective. Are the right team members aware of these objectives?

1.4   Overall payment status:

a. Invoices status; are invoices being submitted on time?

b. What is the amount due for payment, and what is overdue and why? What is the project team doing to expedite the payment from the client?

c. Payment milestones status.

d. If payments are fixed in multiple currencies, how does this compare with the actual mix of currencies in the project cost estimate.

1.5 Profit and cash flow status:

a. Margin or gross margin (overhead and profit) forecast compared with the plan.
b. Cash flow status – if it is negative, when does the project become cash neutral and cash positive?
c. Foreign exchange – Has foreign currency been bought forward? What are the project exchange rates, and what is the variance with actual rates?

2 Scope of Work and Change Orders

2.1

What is the status of client-supplied information that will be relied upon for project execution.

2.2 Changes to scope:

a. Changes identified or requested
b. Changes submitted awaiting approval
c. Changes approved or rejected
d. Cost exposure if work on change orders has commenced ahead of approval. This is a no-no, and the project manager is likely to be reprimanded. Nevertheless, some clients do enforce this in contracts. This situation has to be even more actively managed by the project manager and the implications reported in the monthly report.

2.2.1
Experience shows that the cost of changes is rarely overestimated, and project managers and teams believe they have a greater ability to accommodate changes than is actually the case.

2.2.2
Some contracts may have certain value thresholds, which dictate when change requests must either be started straight away or can allow adjustment to fixed fees. If factors like this exist, then the monthly report needs to identify this status. (See Part III, Section F Contracts, paragraphs 3.7, 3.8 and 3.9.)

3 Project Progress and Status

3.1 Overall project percent complete:

a. Actual percent against plan and compared to the last forecast. This can include performance against early start, late start, contract plan, and internal target plan, depending on company practice.
b. Job to date and incremental monthly progress

c. Main reasons for variance against plan and trends
d. Forecast completion curves and dates

3.1.1

Drill down to a lower level of detail, such as progress by engineering discipline or progress by project area or by process unit. Overall data can mask significant variances and problems at a lower level. (See Part V, Section M 'S' Curves.)

3.2 Resources and Staffing (Home Office and Site):

a. Status of staffing compared to plan
b. Future requirement
c. Critical needs

3.2.1

If additional resources are needed, has the project taken into account the time taken to get these additional resources in planning the future work?

3.3 Technical or engineering status and issues:

a. Key milestones, for example, hazard and operability (HAZOP) review or model reviews
b. Key design parameters, for example, weights and quantities
c. Status of design documentation issued for construction or fabrication
d. Vendor data requirements and status

3.4 Procurement status and issues:

a. Purchase order status (enquired, committed, delivery status, closed out)
b. Equipment delivery status
c. Bulk material delivery status
d. Costs status and trends against budget

3.5 Construction or fabrication status and issues:

a. Construction and other permits status
b. Direct hire progress, productivity, and cost
c. Subcontractors progress, productivity, and cost
d. Subcontractors claims status
e. Cost status and trends against budget
f. Quality control statistics – weld reject rates and so on
g. Labour availability, training, and so on

3.6 Commissioning and plant operations status and issues:

a. Availability of commissioning personnel
b. Number of commissioning/start packs prepared or completed

4   Health, Safety, and Environment

a. Project statistics compared to company targets
b. Leading indicators – accident statistics
c. HS&E programmes/initiatives/training status
d. Safety incentive scheme new initiatives

5   Quality Audits and Status

a. Project set-up, quality, technical and business control audits. Audits have a habit of being put off by project teams due to being too busy. Make sure the project manager schedules them and has them implemented.
b. Engineering office, vendors' work and site.
c. Follow-up and corrective actions.

6   Risk Management

a. Is the risk register maintained and up to date?
b. Risk identification status.
c. Risk memos detailing risk mitigation actions. This should be developed during the proposal phase and then updated during the project. Some 'cold eyes' review may be needed to help the project manager with this. A good approach is to use members of another project to implement this 'cold eyes' review.

7   Client Relations

a. General relationship status. This may be verbal or obtained formally by a scorecard. The project manager needs to be able to provide a summary of the client's opinion or perception as well as their own views.
b. Positioning for future work and projects.
c. Lessons learned. It is important to support future pursuits with the same and different clients and to satisfy third-party quality assessments such as Lloyds insurers, as well as help the company improve its own performance.
d. Claims not covered by the contract change order clause.
e. Other contract issues.

8   Formal Reviews

8.1

There frequently are requirements to provide high level project summaries (commonly called dashboards), which are used at the corporate executive and board meeting level.

Often it is just a sheet or two of key project data. These must be provided in a timely, complete, and accurate manner.

8.2

It is good practice for significant projects (and some 'randomly' chosen others) to have a formal review with a presentation by the project manager and their team. This should take place after about the first six months and then approximately a year later. This can become a significant workload problem for the manager of projects. Delegating this process to the chief executive for major projects helps and smartens up the team. It also becomes a useful mechanism for evaluating the capability of the project manager and their team. Difficult questions that are answered with: "I don't know but will find out and let you know" will be appreciated more than 'waffling'.

9   The Project Management Group

The following two activities give the project management group or department a sense of identity.

9.1

A quarterly project managers' meeting is a useful process for exchanging information, raising internal functional management relationships or conflicts and discussing problems. It also enables the project managers to learn from each other.

9.2

A monthly, bi-monthly, project status description memo (one or two brief paragraphs), covering all projects, helps project managers understand what is going on other projects. In this way they can identify if they are facing similar problems someone else is experiencing and vice versa.

10   Evaluating a Project Manager

10.1

When evaluating a project manager's performance, the project data and performance is only part (albeit a significant part) of the story, and other factors should also be considered.

10.1.1

How is the project performing against the as-sold plan? Is the project performing to plan or even better than plan? If a project is sold with an aggressive cost or schedule, it is possible that even a loss-making project could be a good performance.

10.2   Consider how the project manager interacts with company corporate management.

a. Is the project manager transparent in reporting concerns, issues, and problems? Do they have a tendency to hide project problems and then surprise internal company management or the client when they can no longer hide the problem? It is not uncommon for senior company management to first hear of a problem from the client rather than its own team, which is *never* a good situation.

b. What feedback do you get from a client or your own staff regarding a project manager? Note that complaints about a project manager may not be a bad thing. It may indicate the project manager is doing a good job in protecting the project, so judge the feedback carefully.

c. How does the project manager support other company objectives with their project? Do they support or resist company training requirements, developing people, and providing them with new opportunities? Do they help develop new tools or office capabilities? Ideally, no project should be used as a testing ground for a multiple of new company developments and objectives at the same time. Equally, no project manager should isolate their project from the other needs of the company.

d. What are the project manager's leadership and people management skills like?

11   The Manager of Projects and the Client(s)

11.1

The manager of projects often acts as a corporate sponsor for a project. They interact with a client opposite number at an executive level above the respective project managers.

11.1.1
The biggest danger of this role is that these sponsors start doing the project manager's jobs for them. The manager of projects and their client opposite number need to stay above the day-to-day project issues and activities.

11.2

There should be quarterly sponsors' meetings to review high-level project status, overall business issues and objectives, and so on. The meeting should provide a level of resolution for problems, which could not be reasonably resolved at the project level. Both sets of representatives should be fully briefed and familiar with the current significant issues on the project.

11.3

Issues which can have a significant impact on the outcome of the project or impact the client or contractor's business interests should always be made known to the sponsors.

11.4

The manager of projects role needs to be recognised as one of the key relationship management roles between contractor and client.

Section F   The Owner and Client

There is no doubt that the client or owner as initiator of the project bears a major burden in making the project a success. The key initiatives that are essential and that the client project manager must be deeply involved with are:

- To identify very clearly the scope of work for the project and what, exactly, is to be delivered (see Part IV, Section F Scope).
- To determine the most appropriate contracting strategy to be used to effectively deliver the scope of work for the project (see Part III Section B Contracting Strategy Considerations).
- To pick the team, design the organizational structure, and identify the skills required to manage the contracts selected (see Part V, Section Q, subsection1, Selecting the Team).
- The project manager then has to provide the leadership and guidance, which will allow the selected team to deliver the project safely, on schedule, and at the lowest possible cost (see Part VI, Section B Leadership and the earlier Section D on The Project Management Role).

1   Some Fundamentals

1.1

Ensure that the contractor's (and your own) safety culture is aligned to your expectations and actively maintained at all times.

1.2

Agree on the level of contractor resources to be provided. Put penalties in place to ensure key personnel are maintained.

1.3

Clearly identify all deliverables and exactly what form they will take.

1.4

Document control and project reporting are boring subjects to most people but critical to project success. Agree exactly how these will be managed, produce a master document register and determine distribution early in the game.

1.5

Be clear about any special client-imposed studies and when they should be carried out. These might include hazard and operability studies, hazard identification studies (HAZIDS), and project safety and environmental reviews (PHSERS).

1.6

Ensure that operation's (the user's) input is available from the beginning. Get operations personnel in the project team on a full-time basis.

1.7

Where design contractors are concerned, do constructability reviews. Have the construction contractor review the design for 'constructability'. This is very important when you consider that 10 to 15 per cent of a project's budget is in the design cost and 40 per cent to 50 per cent is in construction costs. Thus, it is essential that construction efficiency is at the forefront of the project manager's thoughts. This also reduces construction changes and variation requests.

1.8

A very clear statement of requirements (comprising the scope and the deliverables) must be developed at the beginning of the project.

1.9

A detailed basis of design must also be developed. At this point it is essential to make clear the requirements for standards, codes of practice, and any special owner company requirements in addition to the normal codes. For example, most experienced clients will have their own design practices that have to be used on all their projects. This is in order to facilitate consistent spares and maintenance procedures.

1.10

There must also be a clear understanding of any regulator's requirements, for example: the health and safety executive, Lloyds insurance, and so on. In other countries, say Norway, there is the Petroleum Safety Authority, Petroleum Directorate, and so on. There will be something similar in other countries.

1.11

Make sure you understand exactly what is in the contractor's overheads and what is in the preliminaries. Also, what the ratio is for productive work to nonproductive work.

1.12

Client team members must actively think about the impact of everything they do and evaluate every action in terms of its effect on the money on the bottom line.

1.13

Authority levels, together with roles and responsibilities, must be clearly understood and approved. Individual team members must recognize and comprehend what it means to be professional in whatever role they fulfil, be it an electrical engineer or an accountant.

1.14

Lines of reporting and communication must also be agreed and understood by all concerned. The client team members must listen and learn to hear what is not being said.

2   Cost and Planning

2.1

The cost estimate must be of a high standard prior to project sanction. Most clients normally aim for a + or −15 per cent accuracy pre-sanction.

2.2

A detailed plan is also needed with a clear understanding of the critical path.

2.3

The planners are the people who should really drive the job. They need to be aware of exactly what time and resources are required for every activity to complete the job and should be on everyone's back to make sure the contractor is performing the right actions at any particular moment to keep the project on track.

2.3.1

Many of the planners are not much more than reporters who tell the client what has been done, rather than what needs to be done. The planner should be like the man who sits on the platform at the end of the Roman galley, beating the rhythm for the rowers to get to where they need to be.

3   Things to Watch

3.1

The quality discipline comprises both quality assurance (QA) and quality control (QC). Client personnel can produce new processes like no one else on earth, and they are good at it (even though there is far too much of it). This is the QA piece.

3.2

The QC part is about the application of the QA procedures, and this is the part that clients tend to fail on. There are numerous examples both in manufacture of materials, equipment, and in the fabrication or construction yards of poor quality. The client must tackle this problem. Project managers must have QC very high on their radar.

3.3

Do not place orders for major equipment packages too early. Design development must be sufficiently progressed to allow orders to be placed without the fear of many changes.

These late changes have a serious impact on the construction programme because of late delivery, late vendor data, and so on.

3.4

Do not start construction too early (before the contractor is ready), which again results in changes and variation orders.

3.5

There always seems to be pressure to place orders and start cutting steel (particularly in the offshore industry) as soon as possible. Project managers must have the courage to hold their fire until they can really 'see the whites of the eyes', so to speak.

3.6

It is worth repeating: do not place long lead equipment orders too soon and do not start constructing too soon because both these things lead to delay and extra cost.

3.7

A client project manager, when under pressure, may want to take over the project execution manager's role!

4   Most Important of All – Safety

4.1

This really has to be at the forefront of the project manager's thoughts at all times.

4.2

Good safety is good business.

4.3

A safe team becomes happy team, a happy team becomes a confident team, a confident team becomes an efficient team, and an efficient team is good for business.

4.4

Poor safety (as some client's know well) wrecks lives, is bad for reputation and brand, damages the revenue stream, and costs extra capital.

4.5

Bad safety is bad for business.

Section G   Achieving Success

In defining success, one has to distinguish between success of the project and success of the project management function.

In any company with an established project management culture, the success of the project management process will be measured by beating or meeting the project objectives of safety, quality, time and cost (with a happy client). "If you can get safety, quality and schedule right, then cost (and reputation) is largely taken care of."[18] Success of the project will focus on meeting the financial returns and benefits. In other words, does the project perform as expected? From a contractor's perspective, they will want to have made a profit and, in the longer term, have established a good relationship with the client for repeat work. You are most likely to achieve this if you start the job right and do the job right. Nevertheless, owing to the diversity of the views of stakeholders, one is unlikely to satisfy all the people involved.

Project success will be measured differently, depending on the project phase. In the early phases, there will be strong external pressures (see Figure I.G.1), and success will be judged on financial and cost issues. As the project moves into the execution phases (the most sensitive to failure issues), the project and project manager will be judged on meeting time deadlines. Finally at the end of the project, commissioning and setting to work, it will be all about quality. At start up, it either works or it doesn't. Unfortunately, the quality element should have been built in to the early execution stages when the project is under pressure to meet time deadlines.

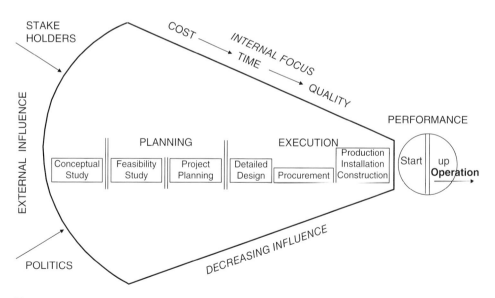

Figure I.G.1

If a project is cancelled or terminated prematurely, then success will be measured by the effectiveness and efficiency of how well the project is closed down: terminating

---

18 Patrick McHugh, see the Acknowledgements.

expenditure, cancelling purchases, reducing manpower and stopping man hour bookings, and so on.

In January 2000 I wrote a paper ('Disconnected Project Management') for the International Project Management Association 15th World Congress on Project Management, held in London in May of that year. In that paper I wrote:

> Project Management does not always work. Data often quoted by the Industrial Society indicates that 77 per cent of projects in the UK fail. In the U.S. this figure is worse at 83 per cent. For specific sectors the statistics are horrific: only 7 per cent of business process redesign projects and 3 per cent of IT projects succeed. Further, 80 per cent of companies that failed had no Project Management infrastructure. The Industrial Society quotes the following reasons for project failure:

Inadequate definition	Inappropriate team
Poor or no planning	Ineffective controls
Wrong leader	Poor communication.
Scope not defined	Unrealistic timescale

> These are similar, but different, issues to many other lists.[19] For example, The Business Round Table in the U.S. states that the poor definition of project scope is the primary cause of cost over-runs followed by the loss of control during design and execution.
>
> By contrast, W.Belassi and O.I.Tukel quote seven lists of critical success factors from .... [leading authors on the subject over almost a thirty year period]. Space does not permit providing details of these lists but, broadly, they are the positive and opposites of the causes of failure. The interesting feature of the lists is that there is almost no commonality of issues. Out of a total of 61 issues only 3 appeared in four of the lists and only one was mentioned specifically 5 times. Allowing for the different terminology, this issue - project controls, could be said to have 10 mentions (but despite this did not appear in all of the lists). The author's own primary prerequisite for project success namely: 'Alignment of Objectives' is only mentioned three times in a variety of forms as: 'definition of clear goals and objectives.' Anyone who has played any version of 'The Prisoner's Dilemma' will know how important the 'alignment' part is and that personal objectives will always carry more weight because of the fact that they are in the control of the individual. Thus, whilst the literature has identified the *causes* of project success or failure, it has not satisfactorily explained the *reasons* behind these causes to help us find ways of dealing with the issues.

Having collected lists of success factors from various sources over the years, it is difficult to disagree with any of them. Whilst there is some commonality to the issues in the lists, the emphasis tends to be on the negative side and their description probably reflects the deficiencies and difficulties within the organizations that produced them.

---

19 Survey results (*The Times Raconteur* supplement, 2/8/2015 – source PMI 2015) for the causes of project failure, in the twelve months prior to publication, show little change in the reasons and their diversity.

Analysis of the lists shows that the reasons for success and the causes of failure can all be grouped into one of John Adair's three elements: task, team, or individual. In projects that fail, one of these elements has been exaggerated at the expense of the other two. Alternatively, one has not been given sufficient attention. Despite this, the conclusion is that it is the basic task issues that are the main problem, which are primarily the project manager's responsibility.

I have been fortunate in that I have spent my hands-on project management experience in organizations with a strong project-management culture, namely, good corporate policies, management procedures, regular thorough reviews and having well-developed systems, tools and techniques. There has always been a total commitment to project success of delivering the client's scope within cost and time targets with the right quality of performance and finally, having a happy client. My career, salary, and bonus depended on achieving these targets, and it was usually challenging, interesting, and fun. I have always regarded a strong project management culture as a prerequisite to success.

Research commissioned by the APM[20] resulted in a report identifying five factors: "As a formula for success; get these right and the rest should follow". Namely:

1. Thorough project planning and review (interestingly this factor was least present in the projects surveyed)
2. Clearly specified and recognised goals and objectives
3. Effective governance with clear reporting and regular communication
4. Project professionals forming a competent project team
5. All parties involved must have commitment to project success.[21]

I think that it is fair to say that the APM research was dominated by systems and services – soft projects. My experience has been primarily physical and technological – hard projects. Now that I have an opportunity[22] to review the APM results with my own experience, it is interesting to see the compelling correlation between them and identify that there is no difference between the success factors for soft or hard projects.

In looking at why some projects succeed and some fail, it is assumed that the prerequisites mentioned above, as well as all the basics, are in place. We know what we are supposed to be creating, by when, and for how much, and (if we are a contractor) we have a contract to do so approved by all the lawyers and senior managers and so on. But still, some go well (succeed) and some don't (fail). Why? What are the essential issues? My analysis in the remainder of this section is on trying to understand what is not usually covered elsewhere, namely the practicalities that shape and influence the achievement of the success factors. It is based on my experience as a hands-on project manager and from what I have learnt as a consultant and from teaching project management, including ten years as director of a master's course.

---

20 'Factors in project success', prepared by BMG Research, November 2014 and 'Conditions for project success, APM research report' 2015.
21 The remaining factors are: capable sponsors [clients] play an active role; A secure funding base; supportive organizations; end users and operators are engaged; competent project teams; aligned supply chain; proven methods and tools; and appropriate standards.
22 I have to emphasize that this section on success (apart from these paragraphs on the APM) was written before the APM report was published.

1   The Project Management

1.1

The primary tool for project success is the project management itself. As already indicated in studying the literature on success and failure, all the failure elements are mostly down to how the project management was implemented.

a.   The secret to success of an organization is to break the project down and turn the processes into less-complex types. That is, turn strangers into repeaters, turn repeaters into runners, and turn runners into routines. (See the beginning of Section A).
b.   As already mentioned, nobody manages a large project; you should manage a series of small projects. Use the product breakdown structure and delegate sections to different project leaders.

1.2

Each participating entity in the venture (owner, co-owners, contractor, joint venture partners, etc.) must have a person or persons at board level nominated as a sponsor,[23] who stay for the duration of the project. They must have a specific interest and knowledge of the work, and they must make a significant contribution to the success of the project.

1.3

The structure of the project management function can be a major contributing factor to project failure. At one extreme is a project manager reporting to a director of projects who is managing a complex organization. At the other extreme is a similar structure but, in addition, they are also reporting to a steering committee representing different organizations with different objectives as well as trying to get a user committee or handover board to agree amongst themselves. (See paragraphs 3.4 and 3.5, Involvement of Users, below.)

a.   "In the view of several of the Computerisation of PAYE's [COP] senior management, the Steering Committee was the single most important reason for the success of the project."[24] Until recently COP was the only large government information technology project that had been successful, due to the project manager and how he fulfilled the project management process.
b.   The complexity or the size of the client team influences the communication process and thus the chances of success. Further, some industries and businesses create really complex client/user – contractor interfaces.
c.   Do not underestimate the difficulty of the communication process between you, the contractor, and the client. Add to that the difficulty of communicating between

---

23   The Project Management Institute's 2015 'Pulse of the Profession Study,' confirms that actively engaged sponsors are the top drivers of projects meeting their original goals and business intent.
24   Extract from a case written by Dr Peter W. G. Morris, Major Projects Association, Oxford. See also footnote 13.

the project manager and line management. Do not forget the interface between the project manager and the project team.

d. Experience shows that relatively few businesses understand the roles and responsibilities within the matrix organization. Make sure that the functional managers fulfil their obligations, as part of the matrix organization and accept responsibility for the quality of the technology. Make use of the benefits of project audits.

e. Once construction starts, the role of the project manager can become severely compromised (see Part IV, Section Q Installation & Construction, paragraph 4.2) since construction can take control by being in direct contact with the client. If the project manager wants to maintain control, then they must move to where the client is. This will mean leaving a deputy to finish off work in the home office.

1.4

Naturally, a key element is the project manager. The management style and competence of the project manager will have a big effect on how the team reacts to their leadership. The project manager *must* establish command and trust and inject energy and enthusiasm.

a. Learn to manage upwards. Make sure that your boss clearly understands when you want help and when you do not want them to interfere.

b. Don't hide problems. Tell your boss everything. If you don't, they will talk to their opposite number in the client organization and make decisions that you won't like, through lack of information.

c. If personnel have to be removed, then so should the opposite number in the client team. This is particularly true of the project managers. Otherwise, if only one party is replaced, then the new person is always at a disadvantage.

d. As stated in Section D The Project Management Role, paragraph 2.8.3, a project champion will be needed for internal projects.

1.5

Management may want to have people released for a new prospect or other reasons. Consequently, identify replacements and plan for the second-in-command to take over. If you want to move onwards and upwards, you had better have a fully trained deputy to take over (see Section D, paragraph 2.8.1).

2   Alignment of Objectives and Client-Contractor Relations

2.1

A project that is a war zone will fail. Some conflict is inevitable since contractors and clients do not have common interests – not in the real world anyway. The contract hasn't been written yet that truly merges contractor and client interests. Nevertheless, it must be assumed that at the corporate level there is a common cause for the project to be a success.

2.2

At the project manager level, the client project manager has to demonstrate (to their superiors, of course) that he or she is being tough with the contractor and getting value for money. Conversely, the contractor has to demonstrate (to their superiors) that things are not being given away. However, this can be done professionally without endangering the relationship.

2.3

The project manager must develop a personal relationship with their opposite number. Similarly, the project manager's supervisor must establish a relationship with their opposite number in the client organization. At least at these levels, there can be a common objective to make the project a success. The client project manager is dependent on the contractor for their success, and the contractor needs the client for career enhancement and future business.

2.4

Most individuals will have hidden agendas. As already mentioned the personal objectives will always carry more weight because of the fact that they are in the control of the individual.

2.4.1

Pick a team where the individuals will fulfil their private agenda by completing a specific role or job within a successful project (see Part V, Section Q, subsection 1. Selecting the Team).

2.5

Be sure every team member understands and is committed to the project objectives.

2.6

See Part IV, Section E Client Relations.

3   Involvement of Users

3.1

The term *user* is usually utilized to identify the end-user operators of a facility. However, the term can also be utilized to indicate the next party in the work process chain, who uses the work produced from the previous function. Thus, involvement of the users is necessary to get them to buy into what they will be taking over. For example, manufacturers and the construction people are the users of the design. It is, therefore, crucial that the construction manager is involved in the early design process. See Part IV, Section Q Installation & Construction, paragraph 1.3 and 1.3.1.

3.1.1

For prestigious building projects, the involvement of the users is achieved by having a design competition. The public is then invited to contribute their comments, or alternatively a consultation survey is carried out.

3.1.2

For infrastructure and development projects, it may be necessary to develop an ongoing relationship with the local community. Issue a regular newsletter to demonstrate that you are listening to their concerns.

3.2

Similarly, the commissioning team will need to be involved in the early design stages in order to explain the sequence in which the facility will be set to work.

3.3

Both types of user need to be involved in the early project formulation stage. Using/contracting another party or organization for the execution stage (as is customary) introduces a potential barrier to success.

3.4

For some types of projects (for example, information technology) it is sometimes necessary to create a user committee to represent the views of the various departments and functions that will be operating the system.

3.5

Similarly, for projects with multiple groups of end users, it is necessary to have a handover board made up of representatives from each group. This enables the project management function to defuse complaints from individuals.

3.6

The foregoing paragraphs address involving the users at a management level of the project process. However, the principle also applies at the designer/construction worker level. Problems occur because the person who designs or constructs an item only performs to meet the requirements of their particular task. They don't necessarily care about matching up with or fitting the subsequent step. We try to overcome this problem with combinations of work at the higher level (for example: design, supply, and fabricate; fabricate and install; design and build; and so on), but the problem still exists at the lower trade levels.

3.7

See Section F The Owner and Client, Paragraph 1.6. Also, Part III, Section B Contracting Strategy Considerations, paragraphs 5.7 and 5.13.

4 Get and Build the Right Team with Clear Roles and Responsibilities

4.1

Push for resources – get the right people. Build a team that has the skills, experience, and behaviours required for the job. Your success depends on the teamwork of the people you chose. Without the right people, it will be ten times more difficult.

4.2

The reality is that the functional managers will offload who they can. Get rid of the dead wood. The team must have mutual respect.

4.3

Roles and responsibilities must be clearly defined.

4.3.1

After you have chosen a key team member, explain (on a one-to-one basis) why they were chosen as the right person for the position. Clarify with them your and their understanding of the responsibilities of the role.

4.4

Explain to the individual what your expectations are of them in the project role.

4.5

Find the expert who is the key to the technological issues.

4.6

See Part V, Section Q for Selecting and Building the team. Also see Part IV, Section D, Mobilisation.

5 Clear and Complete Scope Definition

5.1

For internal projects, it is essential to get a clear brief at the start. Software projects need clarity of what is to be achieved. Physical projects need clear and complete scope definition.

5.2

As already stated The Business Round Table in the United States says that the poor definition of project scope is the primary cause of cost overruns.

5.3

Develop the scope using a product breakdown structure and work breakdown structure in a team process.

5.4

Specifications are difficult to write, and technical people need to improve their aptitude with words. Avoid the tendency to use generic language for descriptions. Be more specific. Do not use words that keep options open and allow for improvements as the project develops.

5.5

For clarity, use 'negative specifications' stating what is not included.

5.6

Know the scope of work, and be sure you have all the necessary information, materials, and tools.

5.7

See Part IV, Section F, Scope.

6   Thorough Planning of the Work

6.1

I now think that planning is the overall key to success (and the feasibility stage must be regarded as part of the planning process), firstly, because all the other aspects that I have listed could be said to be encompassed within the holistic concept of planning. Secondly, examination of the evidence indicates that projects that are extensively and thoroughly planned by having an extended planning phase usually start right. If a project starts right, it will more than likely continue successfully.

6.1.1

The Computerisation of PAYE project (mentined in 1.3.a above) carried out a pre-feasibility as well as a full feasibility study that took over two and a half years to complete and 'the feasibility plan contained an implementation plan of unusual detail'.

The Land Speed Record Thrust SSC project achieved its objective in one day but was planned for six years.

Sir Ranulph Fiennes' journey around the earth's polar axis, using only land transport, took three years but was planned for seven years.

6.1.2

The above projects should be contrasted with the United Kingdom's Nimrod Airborne Early Warning System, the AEW project of the 1980s which compromised the feasibility stages in order to save time and consequently was late, overran the budget, and was abandoned.

6.1.3

The problem for commercial projects is the conflict between spending time in the planning phases against the benefits and financial returns to be obtained by finishing earlier.

6.2

Nevertheless, the project manager *must* start out from the assumption that the plan produced in the initial (feasibility study or particularly the tender) stage may well contain dangerous nonsense.

6.3

A project, even at the proposal or feasibility stage, must be planned to a Level 3 network degree of detail.

6.4

Get buy-in to an achievable end date. Review the schedule with all key members of the project team and obtain their agreement to it – preferably in writing. Get their commitment.

6.5

The tricky bit is how to amend the plan, realistically, to what everyone wants. Whatever is done must be written down and incorporated into the contract (and paid for!) – *state that the schedule can be achieved, provided that such and such is done.*

6.6

See Part IV, Section J, Planning and Scheduling.

7   Planning Communications

7.1

Surveys show that communication is the biggest problem in all organizations. This is because they have not been planned. A lawyer will treat an interaction with another party as a project, and they will plan their communication strategy.

7.2

Agree on how each communication mechanism will be used on the project.

7.3

Circulate information religiously and appropriately. Don't clog up people's in-trays (real or virtual) with unnecessary documents.

7.4

See Part VI, Section A Communications and Section E Personal Skills.

8   The Efficiency of the Project Launch Phase

8.1

The challenge is to allow time for planning the launch phase so that overall time is saved. A well-organized project management department will have developed a skeleton project launch programme (many of the issues during the launch phase are common to all projects) in order to save time.

8.2

Persuade the client to issue a letter of intent/instruction (see Part III Section F, Contracts, subsection 1 Starting work) that gives *authorisation* to initiate work and be paid for an agreed list of activities. The list should include pre-ordering/reserving manufacturing capacity for long lead material and equipment items.

8.2.1
Ideally, use members from the winning proposal team to set up the infrastructure of the project whilst the project team is being mobilised.

8.3

Make maximum use of the team's initial motivation and enthusiasm.

8.4

Good team building should enable the straight line portion of the 'progress' curve to be achieved early.

8.5

If the launch phase is effective and efficient, there is a significantly greater chance that the following work will go according to plan.

8.6

Do not use up project float at the front-end. Float is for when it is needed at the end of the project.

8.7

See Part IV, Section A, Project Launch.

9   Change Control

9.1

As already stated The Business Round Table in the United States says that the loss of control during design and execution is the second cause of project cost overruns.

9.2

Changes are inevitable. The business environment will change, and this will impact on the project. However, be rigorous about resisting 'nice-to-have extras' that were eliminated during the initial evaluation of the project's viability (see Section A, paragraph 1.5).

9.3

A *rigorous* and equitable change control procedure is essential for success.

9.3.1

'The management team's approach should be to discourage changes of any sort. Only alterations which are necessary for safety or to make the facility work or essential to facilitate construction should be agreed'.[25] One should add: or which conflict with regulations.

9.3.2

'The great emphasis on change control must be mentioned as key to the success of the Computerisation of PAYE project'.[26]

9.4

See Part IV, Section L, Variations/Changes/Claims.

10   Effective Decision Making

10.1

Project control is achieved by the decisions that the project manager makes.

10.2

Decision-making involves experience and judgement together with pertinent and timely information. The experience and judgement is a function of the capability of the project manager, and the pertinent information will have been determined in setting up the project control system. The timely element is a function of the efficiency of the reporting system.

---

25 Slightly edited extract from 'A case study of the construction of a terephthalic acid plant for Imperial Chemicals Limited, Wilton, UK'. Project Managed by Dr Roy Whittacker. *Construction Management and Economics*, 1983, 1, 57–74.
26 See footnotes 23 and 13.

10.3

Timely information can be provided (to three decimal places!) by crunching mega quantities of data, using artificial intelligence (AI). Just don't lose sight of the rules of thumb that you should be familiar with, and apply to your particular technology.

10.4

An average decision well timed is better than a good decision badly timed.

11 Tackle Things Today – Tomorrow They Will Be Bigger

11.1

Chase progress relentlessly.

11.2

Check costs regularly.

12 Conclusions for Success

> *Success is the ability to go from failure to failure without losing your enthusiasm.*
> Winston Churchill

12.1

 Understandably, we always know more in the future and are, therefore, more often than not perceived to have failed. Consequently, start promoting the success of the project right from the start of the project. Publicise successful milestone achievements.

12.2

We can deduce from all of the above that project management is a demanding discipline. Putting all people's success and failure lists together; we would find that the final compilation would include every aspect of project management. Consequently, if you fail in any aspect, the project is likely to fail. Conversely, to succeed, you have to do everything right!

PART II

Programme Management

Section A   Programme Management – What's in A Name?

Potato – patata, tomato – tamata, project manager – programme manager: just words and pronunciation or a real difference? Ask many practitioners what the difference is and they will answer: "The salary…". So, to a practicing programme manager what is the difference and is it important? Never mind going into what is a programme manager or a portfolio manager.

From the experience of the principal author of this section[1], managing both major projects and major programmes across several industries, the conclusion is that the academic differentiation can be useful. It articulates the fact there is a difference and that different skills and approaches are necessary. However, in reality, projects and programmes are intertwined at several levels, in all but the simplest of situations.

Consider the design and build of a highly complex navy warship. In itself it would seem an easy case to define, being a classic design, manufacture, assemble, commission, and handover project. You start with a concept and end up with a very definable end product – a warship. However, the ship is part of a broader programme of defence capability acquisition that will comprise aircraft carriers, destroyers, frigates, logistics ships and all manner of other associated materiel. These together will form an overall programme, with each element playing a part in the whole. So what seems like a straightforward major project is actually part of a larger programme. Further, given that a new warship itself will be in design and production for some five years, and then operate with a twenty-five-year design life, it in its own right is part of a programme of, in the jargon of the trade, 'incremental capability acquisition'. So according to funding and capability trade-offs, new capabilities will be added to the scope of the build over time. The ship, therefore, incrementally acquires capability from the very start, each period bringing a more capable ship at the end of the period than at the start. Thus, is the ship itself a project or a programme? It becomes clear that there are nested programmes: an overall programme of defence capability at the overall navy level, as well as a programme of activity to incrementally acquire capability in each specific element of the programme (namely, the ship). There is also undoubtedly a major project as well, in terms of the design, build and commission of a warship.

There are similarities and parallels here with a large corporation having branches right across the UK embarking on a programme of continuous improvement (capability acquisition), in order to stay competitive. This might involve projects such as introducing new products, installation and upgrading equipment, office refurbishments, enhancing the marketing capability, overhauling procedures, updating the website for each branch, and so on. Clearly, an overall programme made up of a collection of projects, some of which might be programmes of work in their own right. All of them will be aligned to the long-term strategy of the organization and will focus on long-term outcomes and benefits.

Railway infrastructure provides another perspective. There are two basic components to rail infrastructure. The first is capacity, route, and linespeed improvements (getting more people more quickly to more places), and the second is infrastructure renewal

───────

1 Martin Arter. See acknowledgements for a brief profile.

(keeping what you have reliable and safe). Both require major projects, such as Thameslink or Kings Cross, but also programmes such as track renewals or signalling renewals. Thus, there is a capability acquisition programme for the overall railway including major programmes such as Crossrail. However, Crossrail itself is a programme of incremental capability acquisition. This sounds remarkably familiar, as indeed it is. It is no different in principle to the defense model.

The national programme of track renewals at Network Rail has some unique characteristics. Firstly, there is no end to the programme – literally. It even puts painting the Forth Bridge in the minor league of epics. Track renewals have been conducted as a rolling programme for over a hundred years and will continue as long as the railway operates. This is quite different to the defence environment where there are long programme durations, of the order of five to ten years, but not endless. Secondly, the programme of work spanned the length and breadth of the UK, from North Scotland to Cornwall and from West Wales to Dover. Thirdly, there existed a legacy of internal and supply chain arrangements. Each individually had a rationale and logic at some time, but there was no current guiding mind or coherent rationale. Those were some of the conclusions of interpreting the work involved in the programme.

A programme of the dimensions of Network Rail's cannot be managed as a whole. Therefore, as with large projects, it had to be broken down into smaller chunks; in this case geographically into five regions, with a senior programme manager in charge of each region. This provides ownership of the five regional programmes that can be planned and executed. The next step was to align the activities of Switch and Crossing Renewal (S&C) and Plain Line (PL) track renewal. Historically, these had been separated as they have different characteristics; S&C being more like mini projects with design, manufacturing, installation and commissioning, whereas PL renewals are more of a production line of standard components. The regions are aligned to be accountable for delivering both in an integrated plan, and the supply chain is aligned to the same arrangement. Lastly, an endless programme needs a motivating mechanism. This is achieved by means of a balanced scorecard of key success factors that are evaluated on an annual basis. Thus, the regions compete with each other every year and similarly for the contractors. It also brings clarity as to what is important, i.e. the dimensions of safety, installed quality, volume delivered, impact of the operational railway and cost efficiency.

Those alterations were the translation of the interpretation of the circumstances and environment into tangible changes. The changes brought ownership and effectiveness into a very disparate programme of works; annual key success factors, that drove competition and performance and brought clarity of vision to an otherwise endless rolling programme of works.

1   Programme Management Conclusions

1.1

Conclusion number one for the practicing programme manager – is that it is about shades of grey and in particular, how to manage and succeed in an environment that is shades of grey. The conclusion from being involved in project and programme manage-ment benchmarking is that, for the most part, the only answer to typical benchmarking questions is: 'it depends…'. For example, consider a typical set of questions:

- What is the most effective contracting strategy – for example, fixed price, target cost, fixed fee and so on?
- Do you insource or outsource design?
- Do you employ delivery partners?
- Which construction design and management regulations roles do you fulfill?

In each case the answer will vary; fixed price is often useful in an open market like civil construction for offices or footbridges. However, a framework contract may be more appropriate in a more specialised market like rail signalling works. Insourcing or out-sourcing design, or the use of a delivery partner, is likely to be dependent on in-house capability. The amount of capable management that is available, the volume of activ-ity and the volatility of the workload are the determining factors. At a regional level rail infrastructure is prone to peaks of workload, and it certainly is in the shipbuilding and power businesses. Consequently, depending on the contracting strategy, the roles of each part of the supply chain and the nature of the project, the various CDM roles will be discharged in a variety of ways. So the answer is nearly always – it depends. There is unlikely to be a correct answer; it is more likely to be a best answer in the circumstances.

Thus, there is no universal best, benchmark solution; it all depends on the industry environment and circumstances.

1.2

Conclusion number two – do not employ the strategies you used on your last programme, no matter how successful. Don't do it unless you are certain the new programme you are managing is a carbon copy of the previous one (which is highly unlikely in today's environment).

1.2.1

There is a real need to work from first principles, which itself needs an understanding of the nature of the programme, the customers, the political and industrial context and the capabilities of your company and of the supply chain, not to mention an excellent grasp of what great programme and project management looks like.

2   Summarizing Programme Management

Life is shades of grey, and there are no universal solutions.

That may sound really inadequate, but, in fact, it is profoundly useful. If you can accept this premise, then it avoids the two most common mistakes observed in project and programme management:

- Assuming the contract and technical specification is all that matters
- Applying what worked well last time.

So what does this mean for a programme manager? The conclusion is that a pro-gramme manager needs the skills to be able to fulfill three key roles.

3   Key Roles for a Programme Manager

3.1   The Interpreter

3.1.1
Stage one is to assess and interpret the programme that is being developed. How much of it is a programme and how much of it is a collection of projects? Who are the key stakeholders? What is the political, economic or social environment? What are your capabilities and what is the capability of the supply chain and most importantly, what does success look like?

3.1.2
Interpreting these dimensions is essential to be able to create a compelling and congruent vision for the programme that recognises its various facets. Without this understanding it is unlikely that success will follow. It is much more than understanding the terms and conditions of a contract.

3.2   The Translator

3.2.1
Stage two is to turn the interpretation of the programme into a compelling vision that engages all the key stakeholders. In other words, to translate the understanding of the programme into a tangible and clear vision and direction that will be essential to success. Indeed creating key success factors is an essential outcome of this stage.

3.3   The Energy manager

3.3.1
Having understood the programme environment and translated it into a compelling vision, the third key skill is to create the energy (as with projects) in the programme team to deliver an exceptional result. This has two dimensions to it: energy creation and energy direction.

3.3.2
Harnessing the energies of all the programme team is essential, tapping into their motivations and creating an energized environment so everyone is giving of their best. Equally important is directing that energy to the end goal and not to the many distractions and cul-de-sacs that litter most programmes.

3.3.3
Having a highly energized team moving in different directions, or in the wrong direction, is more difficult to correct than a lacklustre team.

Section B   Business Change Programmes

The previous section explained the interweaving of projects and programmes. This section focuses on some of the differences from conventional project management and where business programme management has additional emphasis or needs extra attention.

In the same way that a process or construction project will describe the product to be produced, a programme for business projects will describe a blueprint. The Office of Government Commerce[2] (predecessor to the Crown Commercial Service) describes a blueprint as a detailed vision for an organization, covering what the organization will look like when all the projects (forming part of the programme) are finished and the business transformation or change is complete.

1   Blueprint

1.1

A blueprint will comprise a background history and description of the business issues and problems to be solved. It will identify what is necessary to support the future business in terms of:

a. Processes.
b. Organization and people.
c. Tools and technology.
d. Costs/performance and service levels.

1.2

It is necessary to continually check the blueprint against the approved strategic objectives of the business. If the corporate strategy changes then the programme may also need to change.

2   Programme Organization

2.1

Whilst there can be a 'project board' to act as the client on a project, the client for a business programme will be a sponsoring group with a senior responsible owner (the client project manager) ultimately accountable for the programme and the benefits realization.

2.2

The senior responsible owner will establish and chair the programme board responsible for delivering the programme of projects.

---

2 For a complete and thorough exposition of business change programmes see the OGC's *Managing Successful Programmes*, 2007. ISBN 978 0 11 331040 1.

2.2.1 Other roles and responsibilities on the board can be as follows:

a. A Programme manager who is responsible for delivering the mechanics of the projects; basically a project manager of a portfolio of projects.
b. A Business change manager, in effect the users' representative, with the additional prime responsibility for delivering the benefits of the programme.
c. Project managers of the projects forming the programme.
d. Functional/department managers.
e. A representative of internal and/or external suppliers.

2.2.2
This group is too large. In order to reduce the number of people on the board, make the programme manager responsible for and represent the individual project managers, as well as the supplier representative.

2.2.3
Having a business change manager, with equal rank to the programme manager, means that the senior responsible owner is in fact the true 'programme manager.' The senior responsible owner must take real hands on responsibility (for example, managing stakeholder and functional management interfaces). Otherwise, a fundamental rule of project management of only having one person in charge would be broken.

2.3

The organization structure of the individual project teams will be mainly a Balanced Matrix Structure C, Part I, Section D The Project Management Role, paragraph 2.3. There may be the odd strong matrix 'task force' Structure D. The reason why this type of programme is difficult is the conflict between the 'project work' and the 'day job.'

3 Change Stakeholders

3.1

There is a greater diversity of stakeholders. As a consequence, more emphasis is needed to identify internal and external stakeholders and produce a stakeholder engagement strategy. See Part VI, Section F Politics in Projects, Figure VI.F.1.

3.2

There is a need for a structured communications plan to support the stakeholder engagement strategy to explain the planned changes.

4 Benefits Realization

4.1

The benefits to be achieved by the programme require a distinct realization plan that contributes to strategic corporate objectives. This realization plan will involve early communication of the changes that will occur.

4.2

Benefits may be realized after the programme has been completed.

4.3

It will be necessary to stop people using the old systems and working procedures.

5   Gate Reviews

5.1

Part I, Section B Project Management Characteristics, subsection 3 Key Management Decisions, identifies the decision gates that have to be satisfied before a project can proceed to the next stage.

5.2

In change programmes there are more gates and a very formal gate review process. Reviews will focus on the realization of the benefits and assessing that the benefits are balanced with the risks.

5.3

The reviews are carried out by the project board and different levels of approval sought, depending on the capital investment and levels of risk.

6   Project Controls

6.1

The cost, schedule, and resource planning services together with monitoring and control are provided through a programme office (responsible to the programme manager) for all of the projects forming part of the programme. The office also acts as an information and communication hub.

6.2

Documentation requiring additional emphasis or differing from that normally required on a project is as follows:

a. A vision statement forming part of the programme brief.
b. A benefits map showing the relationship between benefits, (equivalent to a product breakdown structure).
c. A list describing each benefit (the deliverables).
d. A strategy and plan for realizing the benefits (the execution plan).
e. The blueprint (the client's brief, business objectives).

f. A project's dossier listing all the projects and showing their interdependency and cross referenced to the benefits map.

g. The stake holder engagement strategy (the client's contracting strategy).

7   Terminating the Programme

7.1

Because business-change programmes can last for many years, people can lose interest. In addition, there is natural staff turnover, and personnel changes will be necessary. Ultimately, the programme manager may not stay to the end because they are needed for the next special assignment.

7.2

Because some benefits have been realized, corporate commitment and the business case will be weakened. Consequently, it is sensible to terminate the formal programme. A few remaining projects may be selected for reassignment to appropriate parts of the business for completion.

Section C Management of Portfolios

To manage a portfolio of projects you need an *air traffic control system*. Any airport (company) will have a range of aircraft (projects) of different sizes, capability, and capacity. Small ones will be buzzing around on routine activities. Medium-sized aircraft will be doing the boring repeat shuttled trips. There will also be a few larger aircraft, which are the more interesting ones to fly, but being unusual they need a lot of support. However, you can't have them all flying around doing their own thing because accidents will happen and disaster ensues. Like projects, most people prefer their flights to be uneventful.

Air traffic control (project portfolio management) is needed to track and monitor the pilots (project managers) and define the air corridors and flight rules (methodologies and project procedures) to be used. This ensures that flights are correctly prioritized, pre-flight checks (feasibility studies) carried out, and collisions avoided so that whole programmes can be coordinated. The air traffic controllers aim to launch and then successfully land the right flights on time and in budget so that the key stakeholders are satisfied. Sometimes air traffic controllers look after several flights at once and, similarly, some managers will often find themselves members of several steering groups.

All aircraft, big or small, need a destination (an objective), passengers (stakeholders), and a payload of goods (benefits) to be delivered. Further, there must be a business case; all flights must add value to the airport's business objectives. All flights need to plan their route and file a flight plan. Finally, all aircraft will need permission to take off (project launch).

Some airlines (departments) may be allowed discretion to do their own thing as long as they steer clear of controlled air space. Similarly, smaller aircraft can do what they like, where complex tracking and coordination systems are not required. However, they must register a flight plan and will have to conform to procedures when taking off and landing (handover). They must conform to some of the rules, particularly if they want help landing or when they get into trouble.

The control tower staff (portfolio/project office) will carry out a capability check. They will ensure that the airport has the right competencies to launch a flight. It will estimate the fuel (budget) required for each airplane and the number of crew (team members) involved. Further, they will keep a log of the payload for each flight.

The 80/20 rule says that you carry the maximum payload and hence, make the most money, out of large aircraft.[3] Thus, air traffic control may amalgamate some flights for more efficient use of fuel, numbers of crew, and delivery of goods. If there are too many small or medium aircraft, there may not be enough crew for the larger aircraft.

Large aircraft are high risk. Consequently, more and more thorough pre-flight checks need to be done before launching the larger aircraft. The pre-flight checks reduce the probability of failure, but if they do crash, the impact is high and it can bring down the business. Since there are many more small aircraft, there is a higher probability of accidents but, fortunately, the impact will be relatively lower.

Air traffic control (portfolio management) will provide a weather forecast of the risks involved and advise the crew on the actions to be taken. Pilots study weather forecasts

---

3 This was the conclusion from a Cranfield MSc dissertation. The company concerned made their money out of large projects. The small projects were there to keep people busy and provide training and experience in project management.

in detail to ensure that the risks are identified, avoided, and controlled to provide as smooth a flight as possible. Further, if you fly into a storm, the ground and senior staff in the control tower are available to give you advice.

Sensibly, no pilot is allowed to fly unless they have done the necessary hours on a flight simulator (managing a smaller project). Naturally, pilots go to great lengths to ensure that their flights go smoothly. As on all projects, project management skills training for the project manager is essential. The crew, with specific roles and responsibilities, will also have attended appropriate training courses. They will have participated in team building and will have learnt to communicate effectively.

On the small aircraft doing the routine activities, there may only be a pilot and co-pilot to do all the jobs – they have to be more multi-disciplinary. With larger aircraft the crew have specialised roles and responsibilities.

The pilot of the aircraft will review the status of the flight compared to the flight plan and advise the passengers (stakeholders) accordingly. However, most of the time the pilot will find that flying the aircraft is boring. It is when the aircraft is hit by a storm that it gets exciting, and the aircraft might be in trouble; it is then that all of the pilot's skills will be needed.

At regular intervals air traffic control updates the plots on their computer screens to see if everyone is sticking to their flight plan. Also, at regular intervals the chief executive of the airport will visit the control tower to review the general status and examine some flights in detail.

1

Portfolio management sometimes includes (major) programmes and major projects. However, I would exclude these two categories and manage them as stand-alone endeavours in their own right.

It is the multi-project programmes of medium and small projects that should be managed by a portfolio manager with the support of a portfolio/project office. Here are some suggestions, in order to maximise the throughput of projects in this multi-project environment:

1.1

Prioritize the projects in order of importance and in the best interests of the company; namely: in accordance with the benefits produced by the deliverables for the least cost and in line with the corporate objectives. Plot 'strategic importance/alignment' against 'capability to implement' and indicate the value of the deliverable by small, medium, and large symbols on the matrix.

1.2

Stop projects that do not produce realistic benefits. Identify the two categories:

a. Short-term projects to solve problems and keep the business running.
b. Longer-term issues, in accordance with the corporate mission statement that will develop the business and add value.

1.3

Standardize the components within the project deliverables, so as to turn 'strangers' into 'repeaters' and 'repeaters' into 'routines.'

1.4

Group the projects into batches of like projects. A manager's span of control (the number of direct reports of different disciplines that can be managed effectively) is of the order of five to seven. The span of control can be increased if the individuals are doing similar jobs (for example: area sales representatives). On this basis an experienced project manager might be able to supervise a larger number of project leaders.

1.5

A portfolio/project office is needed to collect information and data about the various projects. In a multi-project environment, good standardised reporting and forecasting systems are essential. Naturally, it will also provide conventional project management support functions. However, its prime function will be to turn the conventional project management process inside out and manage the project portfolio from a purely resource perspective. The intent will be to optimise the number of people across the maximum number of projects.

1.6

The portfolio manager presents the data to the management board and persuades them to make the right decisions. Decision-making depends on pertinent and timely data. However, owing to the large number of projects and the resource management focus, portfolio management decision-making also involves understanding a significant amount of statistical analysis.

1.7

At each gate review, check that the investment justification is still valid. Check that the benefits are being achieved and see if any projects should be reshaped or combined.

1.8

With an air traffic control system, there are fewer accidents to the business.

2

It is interesting to note that the medical world's multi-project environment implements some of the above. Could improvements be made by implementing more of the principles of multi-project management? Start with the fundamental requirement of at least one person (preferably a few) totally dedicated to each project.

PART III

Feasibility and Contracting

Effective Project Management: Guidance and Checklists for Engineering and Construction,
First Edition. Garth G.F. Ward.
© 2018 John Wiley & Sons Ltd. Published 2018 by John Wiley & Sons Ltd.

Section A   Feasibility Studies

Proper preparation prevents poor performance

Anon.

The principal reasons *why* projects are undertaken (with some generic examples) are:

Financial:	Oil refinery, Channel Tunnel, business change
Social:	Schools, hospitals, and so on
Strategic:	Enter a new market; military equipment
Legislation:	Environmental and safety laws
Political:	Millennium Dome; Scottish Parliament.

Whilst a single project may have elements of all of the above, usually one reason is overriding and will decide the project's future. Most projects that are the focus of this book will be about making money. Financial appraisal is relatively straightforward, and a detailed explanation is given in Part V, Section H. The remaining four involve weighing up morals, values, social attitudes, and so on; this makes their assessment both subjective and difficult.

The owner's decisions prior to a feasibility study for a particular project and the decision to locate in a particular country are not considered in this book. However, a country risk assessment is covered in Part V, Section M Risk and Risk List, subsection 5. In addition, the basis of this section is that the owner does not have the resources to perform a full feasibility study, using their own resources. Consequently, they have employed a contractor to carry out the study in conjunction with their own personnel.

A feasibility study is carried out following a positive outcome of prefeasibility assessments or evaluations. These prefeasibility studies reduce the number of project opportunities down to one or two chosen to study in depth. See Part I, Section G Achieving Success, paragraph 6.1.1.

A feasibility study is a rigorous evaluation of the technical and commercial viability of a prospect. Essentially it includes defining exactly the envelope of variables within which the prospect is profitable. Its objective is to produce a technical scheme that will achieve investor sanction through a satisfactory rate of return on their investment.

Consequently, the feasibility study should identify the options on which decisions have to be made in order to achieve the highest probability of meeting the project or owner's objectives. The options will be studied in a value management/engineering workshop (see Part V, Section S Value Management).

The objectives, constraints, priorities, opportunities, and so on that must be taken into account in order to carry out the detailed design, procurement, and installation/construction of the project need to be evaluated. As such, the feasibility phase can take a number of different forms. As mentioned above, a conventional study defines the envelope of the variables; other mechanisms that fulfil a similar function are building models, prototypes, architectural competitions, and trials.

In order to get the timing of the front-end decisions correct, it is important to understand the characteristic behaviour of projects (see Part I, Section A).

1   Feasibility Study Plan

1.1

Effective and detailed planning influences the time taken to perform any series of inter-related tasks. This is particularly true of the front-end phase since it is always on the critical path.

1.1.1
Develop a plan for the work to be performed by the feasibility study. The feasibility study stage of any project should be planned and treated as a project in its own right with the usual concept, planning, and execution phases. See Part IV, Section A Project Launch.

1.2

Check the request for the appropriate approval authority. Idle hands in the company like to show that they are busy and create a project for themselves.

1.3

Appoint a study leader. Should this be a creative person who might get carried away by the beauty of the technology? Or, should it be someone with good project manage-ment skills but who might not have credibility with the study team? Finding someone with both sets of skills is difficult.

1.4

The feasibility phase is characterised by the need for creativity. 'Creativity is intelligence having fun',[1] and the plan for this phase should support this objective.

1.5

Since this is a project in its own right, the scope of the feasibility study must be defined. As part of the scope definition, identify the deliverables for the feasibility study.

1.6

One of the deliverables will be a feasibility study report. Consequently, determine the format for the report.

1.7

Assess the impact of the project relative to corporate strategy. Does the proposed project support the strategy?

1.8

Estimate the resources, budget, schedule, and effort required in order to produce the feasibility report.

---

1 Albert Einstein.

1.9

Make recommendations for the internal organization of the study or for external assistance.

1.10

Obtain approval to proceed with the feasibility study and allocate a job number and cost code.

1.11

Obtain your terms of reference, or define the terms of reference of the study leader. Agree the degree and level of reporting.

1.12

Allocate roles and task responsibilities.

2   Defining the Project

2.1

Failure to define the project scope at the front end is one of the most common causes for projects failing to meet their objectives, so don't be another statistic. Decide on the division of work. Break the project into its natural smaller packages. Produce a product breakdown structure. See Part IV, Section F Scope.

2.2

Develop the project specification and scope in sufficient detail in order to obtain project approval and the release of funds.

2.3

Defining the envelope of variables, within which the project is profitable, must be done precisely and comprehensively. This cannot be overstressed.

2.4

The impact of, and sensitivity to, each and every variable and combinations thereof must be calculated with the greatest accuracy possible from the available data.

3   The Feasibility Report

3.1

The development of the project is dominated by decision-making. Consequently, it is important to have clear project objectives that will drive this decision-making process.

This may well be fudged because of the classic conflict between cost, time, quality, and safety. See Part I, Section B Project Management Characteristics, paragraph 2.1.

3.2

The feasibility report (see Part VI, Section L Report Writing) will involve and contain most of the following:

a. Description/definition of project (see paragraph 2 above):
   i. Location and layout options.
   ii. A Product and (an initial) Work Breakdown Structure, within the constraints of time and budget. Do not over-develop the detail at this stage.
   iii. Consider the need for sub-projects.
   iv. Identify deliverables for each phase.
   v. What product volumes can be expected?
   vi. Is it proven technology? How reliable is the technology? Is new technology involved?
b. Reasons for the project:
   i. What would be the impact on other projects?
   ii. Identification of alternatives.
c. Interfaces and stakeholders.
d. Objectives of project: (see paragraph 3.1 above.)
   i. User requirements.
   ii. Cost, time, and quality.
   iii. Success criteria.
e. Evaluation of design and cost data from:
   i. What is currently done?
   ii. Historical reports.
   iii. Preliminary surveys – ground, climatic, labour and other resources.
   iv. Physical, mathematical, and financial modelling.
f. Identify the legal implications:
   i. What approvals/permits/licences are required?
   ii. Statutory and planning constraints.
g. The financial model: (see Part V, Section H Financial Appraisal.)
   i. The capital cost estimate with labour costs.
   ii. Net present value of the project revenue or payment terms.
   iii. Rate of return on capital.
   iv. Savings achieved through value management/engineering.
   v. Cost benefit analysis. What are the costs of the various alternatives?
   vi. Impact of: inflation, interest rates, exchange rates and taxatio.
   vii. Operating and maintenance costs.
   viii. A financing plan with sources and cost of all project funds.
   ix. Is the finance stable, consistent and reliable?
h. Risk Analysis: (see Part V, Section M, Risk and Risk List)
   i. Carry out a qualitative risk analysis and where possible identify quantitative data for a sensitivity analysis.
   ii. Perform a stakeholder analysis as a specialised part of the risk analysis. (See Part VI, Section F Politics in Projects.)

    iii. Identify management options for key risks.
  i. Health and Safety issues.
  j. Environmental impact/study:
    i. Impact of: inputs, process, outputs and waste.
    ii. Impact of construction works.
  k. Identified personnel and resource requirements:
    i. What training is required?
    ii. What IT hardware is required?
    iii. Is additional office space needed?
    iv. What equipment and facilities are required?
  l. Recommended contingencies.
  m. Terms of reference for implementation phase:
    i. Frequency of internal reviews.
    ii. Recommend financial approval levels.
  n. Produce summary documentation:
    i. Make recommendations.
    ii. Provide options.

3.3

The end of the feasibility phase is characterised by a major decision point. Whether the project should be cancelled or given the go ahead by the release of funds for the implementation phase. So, make your recommendations clear.

3.4

Make sure that the study job number and cost code are closed out regardless of the outcome of the final decision.

4   Proposed Execution Plan

4.1

Provide sufficient definition of the implementation phase so as to support the project definition in 2.0 above and provide a sound basis for future project control.

4.2

Consideration should be given to making the market place aware that the proposed project is being developed.

4.3

The subjects listed below should be developed in sufficient detail to support the financial modelling but within the constraints of the agreed estimating techniques to be used.

  a. Contract and sub contract strategies – the contract type for the 'Performance of Work':

     i. The allocation of risk.
    ii. Owner involvement/effort.
   iii. Contracting strategy.
  b. Design Options:
     i. In house or subcontract.
    ii. Own design proprietary equipment or standard catalogue equipment.
   iii. Bulk supply materials or subcontract.
  c. Milestone plan. Are there any constraints or deadlines?
  d. Execution plan options:
     i. Cost or schedule driven.
    ii. Risk management plan.
   iii. Availability of resources.
   iv. Construction Strategy, direct hire or subcontract.
    v. Logistics and infrastructure plans.
   vi. Recommended organizational structure.
  e. Quality plan.

5   The Next Step

5.1

The deliverables that are produced at the end of the study will be all those needed for the owner/client to invite competitive tenders.

5.3

Assuming that you, the contractor, have not put off the owner/client by continuously claiming for extras for investigation into additional alternatives; you might be invited to submit a tender for the execution phase.

5.2

During the development of the study the owner's/client's personnel will have been evaluating you. They will be deciding: "Can we work with this contractor?" It is, therefore, essential that the feasibility study contractor has a team of people that builds the right relationships with the owner/client.

5.4

You now have a problem. Having performed the feasibility study, you will know all of the problems, the risks and design details involved. As a consequence, your tender price will be realistic and is likely to be on the high side. The other tenderers, on the other hand, won't know the detail. Consequently, their prices might be on the low side.

5.4.1

It is probably best that a different team of people, who have not been involved in the study, should prepare the tender. See Sections D and E.

Section B   Contracting Strategy Considerations

It's unwise to pay too much, but it's worse to pay too little. When you pay too much, you lose a little money – that's all. When you pay too little, you sometimes lose everything because the thing you bought was incapable of doing the thing it was bought to do. The common law of business balance prohibits paying a little and getting a lot; it can't be done. If you deal with the lowest bidder, it is well to add something for the risk you run, and if you do that, you will have enough to pay for something better.

John Ruskin, 1819–1900

In developing the contracting strategy[2] for a project the following questions and issues need to be understood or addressed:

1   Business Strategy and Stakeholder Alignment

1.1

What is dictated by the ultimate client and is, therefore, a given? (The operator is not always the client. In many countries, the client may be a state-owned entity).

1.2

What are the objectives, motives, needs, and opinions of each of the stakeholders?

1.3

What are the client's minimum conditions of satisfaction?

1.4

The client's key drivers are cost, schedule, new market or new country entry, and so on.

1.5

What is anticipated by all other stakeholders, such as the non-governmental organizations, company shareholders, local communities, and so on?

2   Regional and Local Factors

2.1

What are the key features, and what is the general environment of the area in which the work will be carried out? That is, is there deep water, pristine land, political instability, lack of infrastructure or local supply facilities, and so on?

---

2 See *The Project Manager's Guide to Purchasing, Contracting for Goods and Services* by Garth Ward. Published by Gower 2008. Chapter 3 Contract Strategies, Chapter 4 Contract Categories, and Chapter 5 More About Contracting.

2.2

Have we worked in this area before, and which contractors are familiar with it? Is this a mature contracting and operating environment or not?

2.3

What is our understanding of requirements for applications, permits, and approvals from federal, state, and local authorities? Are the environmental and social requirements clear, and what do we have to do to satisfy internal or external authorities?

2.4

As the client, are we going to, or be allowed to, distinguish between cost and value in evaluating tenders?

3   Market Intelligence

3.1

Do we have sufficient market intelligence about industry capacity, contractor workloads, and so on?

3.2

Have we looked at supply-chain management and access to global framework agreements and supplier relationships? What will be the impact on the project strategy and the contract and procurement strategy?

4   Prequalification Processes[3]

4.1

Are we obtaining competitive tenders, single sourcing (obtaining only one tender from a selection of possible sources) or sole sourcing (only one tenderer is available)?

4.2

What resources are available to manage the contracts? What is the quality of the people, and how many are available? What is their background and culture?

---

3 The subject of prequalification is dealt with in detail in 'The Project Manager's Guide to Purchasing, Contracting for Goods and Services, by Garth Ward. Published by Gower 2008. Chapter 8, Selecting the Tenderers.

4.3

Do we know of all the contractors and or manufacturers capable and competent to do the work?

4.4

Have we established the contractor's interest in and availability to do the work?

4.5

Is there alignment of objectives, and could we enter into an alliance.

4.6

If partnering or alliancing arrangements are to be considered, sufficient time must be allowed to understand the characteristics, compatibility and abilities of the alliancing parties. Forced marriages rarely work well. Are the company's cultures compatible? Are they all going to be familiar with common codes, standards, and practices? Are they really going to work as a team when things go wrong? See Part V, Section J Joint Associations.

4.6.1
The two key points to bear in mind are:

a. Use plenty of time up front to get the relationships right before committing.
b. Do not assume that a successful arrangement on a previous job can just be trans-ferred to a new project and expect it to work the same way again.

5   General Contracting Issues

5.1

Consider paying contractors if they are to participate in design competitions or re-tenders.

5.2

When contractors are grouped into arrangements such as:

Engineering procurement and construction (EPC)
Engineering procurement construction and commissioning (EPCC)
Engineering procurement construction and management (EPCM)
Engineering procurement construction and installation (EPCI), and so on
Give consideration to which contractor should be the lead contractor. See 5.12
    following.

5.3

The lead contractor should ideally be the one with whom the largest amount of money is to be spent. This gives the client more effective and direct control over the financial outcome.

5.4

Do we understand and have the ability or resources to manage the major interfaces being created by the chosen strategy? Has this been adequately considered in arriving at the breakdown of work?

5.5

For large projects that are to be executed in a variety of locations, for example, say, design by a major contractor in London, construction in Korea, procurement from a variety of international manufacturers, all in different places, with different time zones, different languages, and different legal systems, as well as different standards and so on. With so many locations that could be used on the project, can they be minimized or even reduced to just one? If not, do we understand the implications on management resourcing and how we maintain a common standard and focus on a common plan? There is a need to develop coherent and consistent expectations and for a plan to maintain connectivity.

5.6

Generally, construction or installation contracts are of the largest monetary value and, therefore, are the contracts which can have the most impact (good or bad) on budgets and schedules. Consider how these contractors do their work and what they need from the client and others to be able to work efficiently. Equally, is the client able to maximize their influence in design?

5.7

What inputs do they need to make to other contractors' activities so that they may also be efficient? Specifically, get the construction or installation contractors (users) involved in the design process.

5.8

Who is going to be responsible for purchasing, and hence, delivering the supply of material and vendor data to site?

5.9

How are you going to ensure that the design sequence and material supplies are compatible with the fabrication and installation or construction requirements?

5.10

If the client's requirements are not sufficiently developed for contractors to submit realistic prices, consider a two-stage tendering process for the design and build phases. The design team is then novated/transferred to the construction contractor for the construction stage.

5.11

The major confrontation, which traditionally occurs, is that design data and materials are delivered to site late or incomplete. If this issue can be overcome, the constructors have the opportunity to perform and build efficiently, and budgets and schedules will be secured. This assumes they have adequate and quality supervision, good planning, effective materials management and logistics, and a trained competent workforce.

5.11.1

One way to ensure deliveries of critical materials (or the use of a specialist contractor) is for the client to award the contract and then to transfer/novate the contract at the award of the main contract.

5.12

Consider making the constructor the lead contractor. Evaluate any implications of this decision on design and operability. Conventionally, in the past, the design contractor was often the lead contractor who let and managed the construction. When the construction contract is the most expensive, it may be better to be dealing with the contractor that is going to be spending the most money.

5.13

Contracts should allow for construction personnel to be involved with the design contractor to ensure constructability and correct sequencing for the build programme. Further, installation personnel need to be involved with the design of lifting and installation aids.

5.14

Contracts should ensure that maximum onshore commissioning is carried out for offshore projects. For all projects, a commissioning plan and strategy need to be in place and understood. This includes a systems approach, transitioning from the bulk build to systems completion and progressive handover, including handover certification.

5.15

Design contracts should include full details of the required plant operational performance, reliability, spares requirements, and so on.

5.16

Ensure that contractors and manufacturers have good quality assurance processes in place and are also capable of ensuring excellent quality control.

5.17

Understand the lessons and problems from other similar projects and operations.

5.18

Ensure that operations are involved in these contracting strategy deliberations from the very beginning.

5.19

Understand the cost, implications, and benefits of imposing punitive conditions, such as heavy liquidated damages and so on.

5.20

Consider the views and objectives of the potential contractors on all of the above.

5.21

Consult corporate finance for advice on currency arrangements and financial checks with respect to potential contracts.

5.22

What contribution to improved environmental and safety performance can the proposed contractors or contracting strategy make?

5.23

What new technology can be offered?

5.24

In what way can any form of e-business be utilized in either the design, procurement, or construction activities?

5.25

How can the contracting strategy help us to guarantee that everything will work first time and continue to do so? This is particularly important with offshore deep-water or remote installations where remedial work on even a simple, single component can cost a huge amount of money.

5.26

How can the contracting strategy help us to achieve successful start-up? Where start-up means:

a. Meeting first the production date or earlier
b. Ramping up to peak production as planned or quicker
c. Achieving design throughput or more
d. Maintaining planned on-stream time or greater

Section C   Issuing an Enquiry

Look beneath the surface; let not the several quality of a thing nor its worth escape thee.

Meditations VI, 3 by Marcus Aurelius Antonius AD 121–180

For an owner, one of the first steps after project sanction and the development of the contracting strategy is to prepare to issue an enquiry or a request for a proposal (RFP). For a contractor, the first step in the project process is receipt of an enquiry document and the preparation of a tender or proposal.

This phase is preceded by a prequalification process (see Section B subsection 4), whether one is a client or a contractor. The purpose is to find four to six organizations that have the capability to execute the proposed project and who are willing to do so on the terms outlined in the enquiry document.

If necessary, invite expressions of interest and prequalify the names on the list, in order to make sure you have people who have the right background, experience, and expertise.

The following enquiry checklist[4] is designed to be used either by an owner enquiring for a contractor's services or to be used by a contractor enquiring for goods and services.

1   Enquiry Preparation Phase

1.1

An owner proceeding with an enquiry for a main contractor before authorisation would be a mistake. Nevertheless, there is nothing wrong with a contractor getting agreement to proceed with early enquiries for goods and services in order to validate their cost estimate.

1.2

Make sure that all enquiries are properly coded and logged into a register. Check with project controls for the budget and schedule requirements to be assigned to the goods and services to be purchased.

1.3

Perform a risk analysis (See Part V, Section M) with the project team and a stakeholder analysis as part of it (see Part VI, Section F Politics in Projects).

1.3.1
However, the risk analysis for a services contract should already have been performed. It would have formed part of the feasibility study; see Section A.

---

4 The enquiry process is explained in detail in *The Project Manager's Guide to Purchasing, Contracting for Goods and Services*, by Garth Ward. Published by Gower 2008. Chapter 9.

1.4

Are you going to provide a ground conditions survey and accept the risk of unknowns, or are you going to make each tenderer spend time and money doing their own surveys?

1.5

Decide on the appropriate allocation of risk and choose your contract strategy accordingly. Remember the broad brush payment terms labels do not give a true description to the factual nature of a contract – read the contract words.

1.5.1
Decide whether you are going to offer a contract that provides an inbuilt incentive mechanism, say, a target cost contract. Alternately, are you going to offer a stand-alone incentive scheme?[5]

1.6

The procurement group should have either used the corporate database for suitable suppliers and contractors or carried out a market survey. Check where the names for the tender list originated.

1.7

Have preliminary talks with contractors and suppliers before issuing the enquiries. It will help get your project into their work schedule.

1.8

Do not impose more onerous terms on suppliers or subcontractors than your own contract document. This just adds costs.

1.9

Make sure you select the right/appropriate people for the tender list. The wrong/unsuitable organization on the list will submit what appears to be a good offer, and if selected, will give you problems on the project.

1.10

The specification is the key document in an overall contract package. Its importance cannot be overstated. It defines what you wish to buy and, consequently, what the contractor must supply. The specification should be as explicit as possible in defining

----

5 The chief executive of a house-building company said, "Incentives were still required to conclude most transactions." *Financial Times*, 18 Oct 2006.

the scope of work and the standard to which the work must be performed. The specification must define the deliverables required from the service and the type of labour, equipment, and materials to be provided. See Part V, Section A, subsection 2, Inspecting Work.

1.10.1

Whilst this definition document should state clearly what should be done, very rarely does it state what should *not* be done, that is, what is excluded from the project.

1.10.2

Perform a product breakdown structure analysis (see Part IV, Section F Scope) to supplement the written specification of the scope of work or material requisition.

1.11

Collate all the appropriate specifications, standards, and any examples/samples together with all the drawings, sketches, and plans.

1.12

Develop a schedule, showing the project objectives and key programme dates, together with critical items. For an owner initiating a project, this is likely to be a bar chart; whereas, for a contractor buying goods and services, it should be a network or based on a network.

1.13

Decide whether to use a standard form of contract or to use a tailored contract. See Section F Contracts, paragraphs 3.3 and 3.4.

1.14

Check the meaning of key words used in the contract terms.

1.15

Decide on the Incoterms to be used. See Part V, Section H.

1.16

Decide whether the payment terms are to be specified or requested as part of the competitive process.

1.17

Similarly, how is inflation to be handled as part of the competitive process or is a Cost Price Adjustment (CPA) clause to be offered.

1.18

Decide whether to offer a project insurance package, rather than duplicating with the contractor's or supplier's insurances.

2 Tendering Phase

2.1

Decide on the type and method of enquiry:

a. Openly advertised, open or selectively prequalified
b. Local, national or international
c. Verbal, written or sealed tender process
d. A tender to be accepted as submitted
e. A proposal for discussion of options and ideas
f. A quotation (a price) for a standard catalogue item

2.2

Allow a realistic period of time for submission of tenders appropriate to the type and method chosen. Check the schedule with project controls. Give the tenderers time to exercise their creativity. This is the time when money can be saved.

2.3

Tenderers are bound to have questions. Respond to questions and supply written answers so that everyone gets the same information without knowing who asked the question.

2.4

Make arrangements for contractors to visit the project location and respond to their questions at a clarification meeting.

2.4.1
If necessary/relevant tell the tenderers where documents can be viewed.

3 Evaluation Phase

3.1

The analysis seeks to determine the best combination of project execution factors and business terms. The difficult part is selecting the assessment process and evaluation criteria to be used.

3.2

Schedule or set-up a Tender review board.

3.3

Specify the tender opening process, whether a private project process for goods and services or a public opening for contractor selection.

3.4

Specify the duration of the evaluation period and the required evaluation completion date.

3.5

Record the tenders received. Check that the tenders are complete and that there are no missing documents.

3.6

Identify the tender analysis criteria:

a. Technical and execution appraisal
b. Contractual appraisal
c. Financial appraisal

3.7

If all tenders are significantly different from your budget estimate then you should question whether your specification or scope was sufficiently clear.

3.8

Log questions that are required to make the tender compliant with the enquiry.

3.9

Ask tenderers to submit the cost impact of each numbered query and their revised tender price.

3.10

Final post-tender negotiations take place to eliminate queries, minor commercial or technical omissions or qualifications. Identify the 'must haves,' 'like to have and nice to have' issues for negotiation with the two front runners.

3.11

After selection of the qualified tenders, the key to successful contracting is the quality of the proposals submitted and in particular the project team. Interview the key personnel. All other matters being equal, choose the better project manager.

3.12

Approve the recommended supplier or contractor and obtain management and owner approval, if necessary. Be prepared to answer the question: "Why have you chosen this supplier or contractor?"

Section D   To Tender or Not to Tender

The smartest side to take in a bidding war is the losing side.

Warren Buffett

1   The Tendering Decision

1.1

The company should get approximately two weeks notice that an enquiry is on the way. Large projects will have been around and known about for some time. Smaller projects will naturally have a shorter notice period. The purchase of aircraft is advised with a ten-year lead time.

1.2

The three key 'to tender or not to tender' decision makers (see 1.7 below) won't have time to read all the documents involved in a tender request and, therefore, the job of analysing the enquiry is given to an out-of-work project manager.

1.3

Read the request for a proposal enquiry document, and review it with the sales/business development department.

1.4

Check that there is a clause providing an overall limit of liability (see Section F Contracts, Paragraph 3.5 and 3.5.1). The absence of such a clause is a reason for not tendering. This must be negotiated before starting work on the tender. If there isn't one in the client's contract document, and they won't give you one, walk away. You have to ask yourself, 'Why won't they give one?'

1.4.1
Has the client provided a ground conditions survey? If they haven't, ask for it.

1.5

The salesman's job is to bring in the enquiries. Once the enquiry is 'in house', fifty percent of their effort will be directed to selling it internally. They want to be given the credit for a success. For this reason, they should not be a decision-maker in the process of deciding 'to Tender or not to Tender?'.

1.6

Unfortunately, this decision process often takes up to twenty-five percent of the tender period. The company should already have a basic strategy. Therefore, it should not be difficult to identify if and where prospects fit into the business plan.

1.7

The key decision-makers should be:

a. *The managing director or chief executive* – they make the final risk decision. Their neck is on the chopping block.
b. *The business development/marketing director* – they know the marketplace and have a handle on the company tendering portfolio.
c. *The operations/technology director* – they know if the company has the technical capability and if the resources will be available.

Sometimes people want to add in the finance director, but all you need from them is the answer to the questions: Can we afford it, and can we get the finance for it? Similarly, people want to add the legal executive, but their attitude can be that it is far too risky to be in business, anyway. Nevertheless, you will need the legal people to do a risk evaluation on the contract documentation. These two functions have a contribution to make, checking that the key people are being consistent and logical, but they are not necessarily decision-makers.

1.7.1
When I ran a proposals department, I was asked to be part of the decision group. I turned it down on the basis that it was up to key decision makers; my job was to produce a quality proposal.

2   The Tender Decision Analysis

2.1

Considerations for the tendering decision should be:

a. Company strategy
b. Project size
c. Budget and resources

2.2

The key decisions areas will be:

a. Probability of award to the company
b. Ability to do proposal
c. Ability to execute project
d. Probability of the project proceeding

	Criteria	Positive 10 9 8 7	Neutral 6 5 4 3	Negative 2 1 0	Rating
1	Importance	Strategic	Somewhat	Minimal	
2	Qualifications	In-House	Average	New Area	
3	Competition	Sole Source	Unknown	Competitor	
4	Market Intelligence	Good	Up to Date	Little	
5	Client Rapport	Good	Known	Unknown	
6	Proposal Effort	Light	Typical	Heavy	
7	Proposal Timing	Comfortable	Reasonable	Tight	
8	RFP Exceptions	None	Minor	Major	
9	Team Availability	Many	Adequate	Limited	
10	Project Schedule	Comfortable	Reasonable	Tight	
11	Pricing Strategy	Noncritical	Competitive	Must Cut	
12	Profit Margin Potential	High	Normal	Low	
13	Work Location Problems	No Impact	OK	Significant	
14	Risk and Liabilities	Low	Average	High	
15					

Figure III.D.1

2.3

The most effective way to turn a very subjective process into a more objective process is by means of a Tender decision matrix; see Figure III.D.1.

2.3.1
Get the company to agree on the criteria, but limit them so that the matrix fits onto one side of A4 paper. Leave room for comments covering:

a. Features of note
b. Key risks
c. Specialist resources required or restraints
d. A 'completed by' signature. This enables the decision-makers to make their own assessment about any bias or lack of objectivity. As an out-of-work project manager, make a decision whether you want to lead the proposal effort, or don't you like the proposed project and don't want to be the project manager?

2.4

More emphasis has been given to the proposal issues in the matrix, since if you can't be positive about the proposal, you will not win the project. You have a little time to resolve the project issues whilst the tender is being prepared.

2.4.1
Importance: Is it a key prospect in the business plan or has it come in off the street?

2.4.2
Qualifications: does the company have the experience or is it a new area of business?

2.4.3
Competition: is this a single source enquiry that scores a 10, or is it one of ten or more tenderers and as a result it might score a zero?

2.4.4
Market intelligence: when did we know about this project? Has it changed over the time that we have been chasing it?

2.4.5
Client rapport: who are the client's key people, and do we know them? Does anyone know one of them personally? What were they like on a previous project?

2.4.6
Proposal effort: will this enquiry involve a lot of people developing new material, or can it be put together by modifying material from another tender (not really recommended if you want to win)? Have we got the people available to do the work or are they busy on other proposals?

2.4.7
Proposal timing: have we been given sufficient time to develop tailored material focused on the client's objectives, or will we be in a rush and make mistakes under pressure? Do we have time to analyse a network?

2.4.8
RFP exceptions: have we had an enquiry from this client before and come to terms with the way they do business? If we have too many exceptions to the enquiry documents, and in particular the contract, we will be ruled out as *non-compliant*.

2.4.9
Team availability: do we have a ready-made team coming off another project at the right time, or will a team have to be cobbled together from all over the place?

2.5

Project schedule: is the schedule being imposed upon us, or do we have to develop one? A good schedule can be a winning strategy. Can we develop a network to substantiate the schedule?

2.5.1
Pricing strategy: what are the client's evaluation criteria so that we know where to reduce costs? Will there be lots of changes? Are there any comments on the percentages for: overhead, risk contingency, allowance for inflation, and negotiating margin?

2.5.2

Profit margin potential: have we got any innovative ideas for reducing costs that we can share partly by reducing our price but also increasing our margin? Are there areas where we can make additional money on services?

2.5.3

Work location problems: Has a country risk assessment been carried out? See Part V, Section M, subsection 5.

2.5.4

Risk and liabilities: has the legal department had a good look at the contract? Is there a limit of liability clause? What guarantees are required? Are there any unacceptable risks such as 'unknown ground conditions'?

2.6

You may have to carry out a modified analysis if you have to purchase the enquiry documents. For example:[6]

"To obtain the tender documents the bidders will irrevocably deposit the sum of XXX."

"Only the bidders who obtain the tender documents after having paid the amount of XXX have the right to submit bids upon this tender."

2.7

The matrix helps focus where you are good and need to boast. In addition, it also shows where effort will have to be made to overcome a low matrix score/weakness.

3   The Final Tendering Decision

3.1

If the risks are reasonable and you think that they can be managed, if the company has done work for the client before and relationships are good, and if you think the company can provide the best combination of technical and project execution proposals and contractual and commercial terms; go for it and move to Section E Preparing a Proposal for a tender.

3.1.1

I made a persuasive assertion, saying, 'It's our turn, they like us, we know the project, and have some innovative ideas' and so on. We lost and I lost credibility with the executive decision makers for the next six months.

3.2

If any score in the matrix is zero, then don't tender.

---

6 A small sample from the financial press.

3.3

You must have a unique selling feature and be able to produce a professional proposal. If you do not have everything right, one of your competitors certainly will, in which case you may as well not submit a tender.

3.4

You must be able to answer the question: "Why should the client choose us?" If you can't answer this question, don't tender.

3.5

Analyse the decision to tender. If you do not stand a chance or do not want the job, do not waste everyone's time and money.

Section E   Tendering and Proposal Phase

In war there is no second prize for the runner-up.
Omar Bradley, in the *Military Review* February 1950

Successful tenders or proposals have to convince a client of three things:

1. That the company has the necessary experience, resources, and systems.
2. That the company can apply them for the benefit of the client and achieve their objectives better than the competition.
3. That the company has the lowest risk combination of technical and project execution proposals and contractual and commercial terms.

Achieving the first requirement will get the company short listed. This requires good quality standard material that every company can produce. It is achieving the second requirement that will get you to the final. In order to achieve this, the proposal document must:

- Be responsive.
- Stick to the agreed theme for the proposal.
- Highlight what will be provided to the client.
- Highlight why the project should be awarded to the company.
- Clarify or explain what has been written to avoid misunderstandings.
- Illustrate the text, where possible, with charts, graphs, and diagrams.
- Write logically and arrange information in a logical sequence.
- Be project specific. Explain how issues meet the client's objectives.
- Be relevant. If it does not apply to the project, leave it out.
- Be word thrifty. Express the point in as few words as possible.
- Demonstrate understanding of the project, the scope, and any special features or problems.
- Make the proposal document sound as though it was written by the client's own people.

Achieving the third requirement will satisfy the client's evaluation criteria and get the contract.

See Part IV, Section A for launching this process as a project in its own right.

If you have never done business before (doing work or exporting goods) in a specific country, then it is essential that a country risk assessment is carried out (see Part V Section M, subsection 5.) long before any enquiry arrives on a desk and an approval to tender is required (see Section D).

1   Tendering Preliminaries

1.1

Obtain a record of all agreements and discussions held with the client prior to receipt of the enquiry.

1.2

List the business objectives for the proposed tender or proposal in terms of what your company wants to get out of the project.

1.2.1
Remember that the decision to submit a tender may be made on the anticipated return based on the investment of the tendering costs. If the cost of preparing the proposal overruns, then management (if they knew what the final costs were going to be) might not have made the decision to tender.

1.3

Develop the strategy for the performance of the work, highlighting any unusual aspects:

a. Description of facilities, location, and preliminary capital cost estimate.
b. Scope of services and preliminary estimate of their costs.
c. Anticipated project programme and man hour estimate. Suggest improvements.
d. Tender/proposal submission date.
e. Client background, contact details, and any previous tenders.
f. Competitive position and probability of award.
g. Type of contract.
h. What is the contracting position, and what are your responsibilities? Are you the main contractor, the sole contractor, or a subcontractor? Will you need to form a joint venture?
i. What are the technical or commercial risks, liabilities, or guarantees?
j. Are there any exceptions or deviations from the enquiry document?
k. Suggest innovations, alternatives, and improvements in general.

1.4

List the client objectives, that is, what they want to hear from you.

1.4.1
Translate the objectives into cost, time, and quality objectives for the tender or proposal and fix the budget and schedule for the tender or proposal.

1.5

Identify the proposal theme and selling strategy for winning the work. Identify any special emphasis, innovative ideas, alternatives, deviations, or exceptions.

1.6

Write the executive summary and use it as a briefing document for the various groups contributing to the proposal. (See Part VI, Section L Report Writing, paragraph 5.1).

1.7

Arrange and conduct the proposal or tender kick-off meeting:

a. Invitees
b. Agenda
c. Work location
d. Assign team tasks
e. Priority input required
f. Get buy-in to the budget and schedule
g. Get commitment to winning the work
h. Target date for management review

2   Developing the Tender or Proposal – In-house Work

2.1

Decide on the contents of the tender or proposal, and start a dummy book in which to insert completed sections.

2.2

Order covers with the client's name and logo together with project name and your company's name (depending on the style of the manual, this is often on the critical path). Order file dividers.

2.3

Prepare a comprehensive product and work breakdown structure for both:

a. The tender or proposal
b. The project

2.4

From the results of 2.5, compile a thorough and accurate list of deliverables required for both 2.3 a and b.

2.5

Ensure that an accurate and realistic scope of work and services is produced, including all the technical services required.

2.5.1

If the client has not thoroughly and tightly defined the scope, include your product and services breakdown structure, together with your scope document in the tender or proposal.

2.6

Perform a thorough risk analysis with the proposal team. Use the product breakdown structure to carry out the risk analysis and create a risk breakdown structure (see Part V, Section M Risk and Risk List). Clearly list and evaluate all risks (be realistic and do not let people delude themselves that it will be all right when the time comes). Evaluate the risks that can be avoided and controlled. Develop mechanisms and plans for handling the risks. Consider how to address some of the issues in the tender document.

2.6.1
Check whether any unusual risks can be met by insurance.

2.6.2
When developing the risk breakdown structure, look for opportunities for doing things better, faster, or cheaper.

2.7

Prepare the proposed project execution plan, with an emphasis on selling your ideas, bearing in mind that you will have to live with your plan(s) (see Part IV Section C, Getting Organized, Subsection 5), including:

a. A programme – schedule.
b. A mobilisation plan.
c. An organization structure.
d. Identify the skills required for the project team.
e. Description of the project scope supported by the product breakdown structure (paragraph 2.3 b).
f. Execution plans for engineering design and procurement supported by the work breakdown structure. Identify risks, problems, or areas of difficulty and the proposed solutions.
g. Decide on the use of consultants or partners.
h. Identify the fabrication strategy – piece small, pre-assembly, or modularisation.
i. Identify the construction strategy – direct hire or subcontract.
j. Works testing and vendor commissioning.
k. Full load performance testing. Resources and sequence for setting to work/commissioning.
l. Acceptance criteria and handover.
m. Spares policy.

2.8

Prepare a quality assurance plan.

2.9

Prepare safety and environment procedures and plans.

2.10

Prepare a medical plan for remote locations.

2.11

Delegate the tasks and deadlines for each discipline to provide input for the tender/proposal. Be strict so as to avoid last-minute panics.

2.12

List all necessary permits, regulations, and local and national laws that could affect the project.

2.13

Ensure the design team passes accurate data to procurement for pricing and delivery information (if it is only a guess, make them say so).

2.14

Obtain an assessment of the industrial relations situation and potential labour problems.

2.15

Survey the local labour availability for manual, technical, and clerical personnel.

2.16

Check whether any special labour agreements or restrictions are in operation.

2.17

Identify those risks that must be covered by insurance and how much they will cost.

2.18

Determine the overall project liabilities.

2.19

Develop a costing/pricing system for:
a. Data processing
b. Reproduction
c. Communications
d. Records management
e. Computer-Aided Design
f. Control systems

(These items could be a useful source of extra revenue.)

2.20

Develop a project staffing plan.

2.21

Review CVs of proposed key personnel (remember you will only be as good as your team). See Part V, Section Q, subsection 1, for selecting the team.

2.21.1

Tailor the CVs to the project, particularly the achievement of cost and schedule targets. Make them interesting, not just lists of projects or assignments. Describe what was the individual's achievement or contribution. In addition, say what the project achieved that was novel or different.

2.22

Ensure that design, procurement, and fabrication/construction data are provided to the estimating and cost and scheduling departments.

2.23

Review and approve construction scope, organization, schedules, budgets, manpower, equipment, and labour conditions.

2.24

Include testing tolerances and get them agreed by the commissioning people.

2.25

Review and approve project estimates, schedules, and planning data.

2.26

List all technical and commercial exceptions to the enquiry.

2.27

Look for areas of probable/possible future work associated with the project and evaluate the potential extra margin – but be realistic.

2.28

Review and coordinate the input from each group.

2.29

Establish and monitor for the tender/proposal: man hours, cost and schedule, reporting and control systems. Do not overspend to such an extent that it invalidates the decision to submit a tender (see paragraph 1.2.1 above). It is, nevertheless, very difficult to control proposal costs due to time pressure and the focus on being successful.

2.30

Ensure that the tender or proposal is completed on time for the due date. Do not forget to allow time for delivering the tender!

3   Coordinating with Third Parties

3.1

Establish the degree of involvement of other organizations:

a. Partners
b. Joint venture companies
c. Consultants
d. Specialist contractors
e. Specialist vendors
f. Proprietary process owners

3.2

Determine and define the exact split of work between each entity.

3.3

Ensure that all parties are aware of the input they must provide for the tender or proposal and when their input is due.

3.4

Ensure, by involving their top management if necessary, that you have their total commitment to the job.

3.5

Negotiate and agree the split of liabilities, ensuring that the other party accept their equitable share.

4   Coordinating with the Client

4.1

Obtain a list of all client personnel involved with the project. Make contact and establish a good working relationship with each of them.

4.2

Arrange and schedule tender or proposal clarification meeting(s) and invite participants.

4.3

Arrange and schedule visits to the project location and invite participants. This is an opportunity for the team to build relationships with their opposite numbers.

4.4

Develop a public-relations strategy in conjunction with the client.

5   Commercial

5.1

Arrange tender (bid) bonds, lines of credit, and bank guarantees. The enquiry might say: "The tenderers shall enclose a bank guarantee/participation bond, which must not amount to less than 3 per cent of the total value of the tender."[7]

5.2

Review the proposed payment terms, advance payments, fee retention, payment currencies, letters of credit, and cash flow. Make sure that you are not going to be financing the client. The cost of money can make a significant difference to the profit/margin.

5.2.1
Consider whether the cost of tendering should be included in your price.

5.3

Identify any specific requirements from project loans or financing conditions.

5.4

Check whether the client requires financing.

5.5

If finance is required, develop a financing plan in conjunction with banks and funding agencies.

---

7 An example from the financial press.

5.6

Identify the main contract liabilities. Can they be mitigated in subcontract and purchase order contract clauses?

5.7

List all possible costs, such as taxes, duties, insurances, royalties, and licensing.

5.8

Develop a tax strategy and currency plan.

5.9

Develop a pay and benefits plan for the project location and review:

a. Hourly rates
b. Additives
c. Other payroll related costs
d. Fee/margin

5.10

Develop a project cash flow requirement.

5.11

Identify any local content requirements.

5.12

Consider qualifying your tender. Then during negotiations eliminate the qualifications and consequently, reduce the price in order to get the job. However, too many qualifications may not get you to the negotiating table.

6 Reviewing the Tender or Proposal

6.1

Compare this tender or proposal with previous similar projects in order to highlight any unusual or restrictive conditions.

6.2

Review the text of the whole document(s) to ensure that the style is uniform. Does it sound as though the client project team had written it themselves?

6.3

Make sure that illustrations support the text and vice versa and that they can be easily compared with each other.

6.4

A consistent complaint of all project personnel is that the front-end people, trying to win a contract, make promises that certain things are possible and objectives realized, when in reality they may not be achievable.

6.5

Schedule, invite participants, and conduct tender and proposal review meetings:

a. Technical
b. Legal
c. Commercial

Conduct fully detailed reviews and ensure all departments are in full agreement and have them commit to their agreement *in writing.*

6.6

Schedule review(s) with management. Make sure that they receive all the appropriate documents in adequate time before the meeting is scheduled.

6.7

Carry out a detailed review with management and ensure that they are fully aware of:

a. The scope of services
b. The contract terms (and conditions)
c. The risks
d. The potential profit/margin

Remember, management will not have had time to look at the documents in detail, so it is up to *you* to point out *all* potential problems.

6.8

If management cut the costs or schedule in order to win the work, *ensure you have it in writing,* but it might still be your fault!

7 Before Submitting the Tender or Proposal

7.1

Have an expert, knowledgeable proof reader read the tender or proposal.

7.2

Develop a negotiating strategy, including how much you are willing to concede to get the job (but remember you should never give something away without getting something in return).

8 After Completion of the Tender or Proposal

8.1

Continue working on the tender/proposal in order to find any mistakes and risks. These can then be resolved or eliminated during the negotiations.

8.2

Assist the sales or business development department in presenting the tender/proposal to the client.

8.3

Ensure that accurate records are kept of all aspects of the negotiations. However, you may not be present during the negotiation. You need to be available in a back room to give advice and be kept up to date with what the sales people are doing. This prevents you, the project manager, being tarnished with any ill will that may be generated during this process.

8.4

The closer you get to winning the contract, the more you will want it. Thus if, during the negotiations, concessions are made (that is, to cut the cost or schedule) ensure that management is informed and that they agree. *Get their agreement in writing.*

9   Proposal Team Presentation

If you are making a presentation of your proposal, it means that you are in the final few (two or three) being evaluated. The quality, effectiveness, and impact of the presentation, together with the people involved can win the contract.

See Part VI, Section G for presentation skills details. This subsection 9 amplifies the additional requirements and complications of a team presentation of a tender or proposal to a client.

9.1

Presenting a tender or proposal to a client is likely to involve at least five people covering the main disciplines. For any technological project, there will be project management, project controls, design engineering, procurement, and construction. There may also be other experts involved such as safety, industrial relations, commissioning, specialist designers, or shipping co-ordinators. How many of these will depend on the nature of the client's enquiry. Fortunately, the client is most likely to identify who they want to see and what topics they would like included in the presentation. Nonetheless, brief the others just in case the client decides to test you.

9.2

Firm up a running order. Nevertheless, be prepared to change it. A sophisticated client is most likely to ask for changes to the agenda that they sent to you, in order to see how well you can cope with changes.

9.3

The project manager has two roles: to define and manage the strategic plan or agenda for the whole group and to prepare and deliver their own presentation as part of the overall event.

9.4

Your objective is to convince the client that you can provide the client with the greatest confidence that they have chosen the lowest risk combination of technical and project execution proposals and contractual and commercial terms.

9.5

Everyone involved is not just demonstrating how good they are at making presentations, but selling themselves as an excellent manager in their proposed role. In addition, and most important, the team needs to come over as a team and support each other when a member is in difficulty (see 9.18 below).

9.6

The team presentation will require all formats (outlined in Part VI, Section G). You will need to explain how you propose to manage the project (subsection 2). You will need to persuade the client that you have the best possible execution plan (subsection 3) and you will need to express your point of view concerning details of the design (subsection 4).

9.7

Identify an overall theme for the presentation.

9.8

Demonstrate that the company wants the work right from the start, by arranging for the chief executive to introduce the whole process. After the brief introduction, they should leave the room. Do not let them be cornered into answering questions.

9.9

In a presentation of this nature, it is essential to have name badges and name place cards. It helps the client to remember who you are.

9.10

One of the first logistic issues to be addressed is how to hand over from one speaker to the next. Does the first speaker say "I will now hand over to Fred (or Fred Smith?), the procurement manager", and/or does the next speaker say, "Good morning, I am Fred Smith, the procurement manager?"

Perhaps the following works: "I will now hand over to Fred Smith" and the next person says, "Good morning, I am Fred. *I will be* the procurement manager."

9.10.1
It may sound obvious but make sure you know each other's names!

9.11

Brief senior management to be available during the lunch break. Arrange for someone to update them with what has transpired during the presentation and highlight any issues that need massaging.

9.12

Insist on rigorous time keeping from all presenters.

9.13

With so many individual presentations and a client that *will* interrupt or ask questions, you are almost bound to have timing problems. Therefore, it is essential to have identified one or two portions that can be omitted.

9.14

If you are using an experienced senior executive from another part of your company, they will need very firm control by the project manager. Despite this, their ego will probably make them seriously overrun their allocated time. In which case, you will be using your entire drop-out contingency plan.

9.15

There are specific points that you want to make with your presentation; therefore, if you leave something out, you will need to quickly summarise them. Write them up on a flip chart. This also applies to any area where the client challenges you by saying "I don't think we need to go into that." You will need to respond with: "I just wanted to make the point that a, b and c ..."

9.16

Eliminate areas of overlapping content.

9.17

Emphasize that team members must look interested in the speaker – no matter how many times they have heard it in rehearsals.

9.18

The project manager should take the responsibility of accepting all questions and then allocating them to a particular member of the team to answer. If a team member is in difficulty, and someone else knows the answer, they should look at the project manager and say, "Perhaps I can help with my experience of ..."

9.19

This type of presentation needs additional team rehearsals in addition to the individual preparation. Use a red team review process. Find two, or preferably three, experienced colleagues who know little about the project to act as the client. Provide them with the enquiry and proposal documents one or two days beforehand, and then do a full scale dress rehearsal. Get the red team to behave like a client. Get them to ask every awkward question that they can think of. Get them to be difficult and act bored/silent as well as hostile. Be prepared for the client questions below; see subsection 10.

9.20

With a proposal/tender presentation, the client is almost certain to want to interview people other than the presenters. Make sure that you have briefed these key team members (who may not have been involved in preparing the proposal) with the main aspects of the project. Just in case, brief more than one person from each discipline in order to demonstrate experience in depth.

9.21

Make sure that car parking has been organized for your visiting client. Brief the doorman to let you know immediately the client arrives.

10 Possible Client Questions for the Proposal Team

10.1

General: "Is your tender fully compliant with the enquiry document?"

10.2

Chief Executive/MD: "Are you willing to accept the XYZ risk under discussion?"

10.3

Project Manager:

a. "What are your best/most valuable attributes as a project manager?"
b. "What makes you think that you are a good project manager?"
c. "What is your project management philosophy?"
d. "What methodology do you use?"

10.4

Procurement:

a. "Engineering defines the scope, construction determines the delivery, the enquiry cycle is fixed, the client and team define the tender list, so what does procurement bring to the party?"
b. "Do you subscribe to First Point Assessment (or other relevant industry body)?"
c. "Procurement produces many reports; which is the most useful?"

10.5

Engineering/Design:

a. "Don't you think that it is most sensible for engineering to do the purchasing of materials and equipment and then go to the site to supervise the installation of equipment?"
b. "As a designer, designing to a budget, what are the dilemmas in choosing the right materials?"
c. "How do you reconcile the conflict between the functional parameters (the manufacturer's interest), schedule and cost requirements (the contractor's interest), and the operating costs (the user's prime interest)"?

10.6

Construction:

a. "If you are working on another project, how do you provide the necessary constructability input to engineering?"
b. "As a U.S. company, it is our view that there is no such animal as a British construction manager. What are your views?"
c. "What is the role of the project manager during construction?"
d. "Should the project manager be on site during construction, and if so, when should they move to site?"

Section F Contracts

The law is just about the meaning of words and a few simple rules.

David Wright

A man who is his own lawyer has a fool for a client.

Anon.

What the Contract Means
- The most important rule is that the contract, when in writing, as almost all commercial contracts are, is presumed to be a complete and precise statement of exactly what each side has agreed to do so that any failure to do so is a breach of contract.
- It is a breach of contract to prevent or impede the other side in carrying out the contract.
- Everything must be done when stated in the contract or within a reasonable time.
- The contract will always require work or equipment to be of a reasonable level of quality.

How the Contract Is Made
The mechanism used by law to decide whether a contract has been made is that of *offer* and *acceptance*. This means, simply, that for a contract to exist, one party has to offer terms which the other party accepts.

This leads to a number of *rules*:

A. Each party can make and accept offers.

B. The offer can be made in any way: by telephone, letter, e-mail, or across the negotiating table. They are all equally valid.

C. The offer must be complete. It must include:

   i. A description of the equipment or work to be supplied and a timescale

   ii. Price or prices and terms of payment

   iii. Contract terms/conditions

D. The offer must be firm. If it is incomplete or subject to further discussion or 'confirmation', it is not an offer.

E. Until an offer has been made there cannot be a contract because the contract is made by *acceptance of an offer*.
   The result of rules A to E is that the enquiry and all other preliminary documents and exchanges have no contractual effect.

F. The offer ends when it is rejected or at the completion of the validity period. In the commercial world, offers should always have a fixed validity period, thirty days or sixty days, for example. An offer with no fixed validity period will be valid for a reasonable period.
   An offer and a revocation need to be received before they are effective.

G. A counter-offer cancels and replaces the previous offer. Commercial contracts are usually made by an enquiry, an offer, negotiation, some counter-offers, and then acceptance of the last counter-offer.

H. Acceptance can take place in several ways.
Acceptance can be by any explicit statement verbally or in writing, by whatever means or by any act that implies acceptance. Acceptance, if posted, takes effect the moment it is posted.

I. Acceptance must be by the person or organization to whom the offer was addressed or by someone acting as their agent.

J. Acceptance must be unqualified and of the whole of the offer. Qualified or part acceptance will usually be a counter-offer.

K. Acceptance makes the offer into a contract so that the terms of the contract are what was in the offer and the date of the contract is the date of acceptance.

What This Means in Practice

Stage One: Buyer and seller discuss a potential contract.

Stage Two: Buyer sends seller an enquiry. The enquiry describes the work and uses the buyer's standard conditions of purchase.

Stage Three: The parties discuss various queries.

Stage Four: The seller quotes a price for carrying out the work to the buyer. In doing so, he states "Generally, we accept the terms set out in your enquiry but would wish to discuss various points further with you before accepting a contract."

Stage Five: The buyer and seller discuss the points with each other but do not reach final agreement.

Stage Six: The seller now submits a Tender to the buyer, in accordance with the original enquiry and points agreed during Stages three and five. The Tender is subject to the seller's standard conditions of sale.

Stage Seven: The buyer responds by asking for a price reduction of £X and that his conditions of purchase should apply.

Stage Eight: The buyer issues a letter of intent to the seller. The seller commences work.

Stage Nine: The seller offers a price reduction of £Y and agrees the change to the conditions of the contract.

Stage Ten: The buyer issues a purchase order in accordance with Step nine.

Stage Eleven: The seller accepts the purchase order issued immediately by e-mail and a few days later by letter.

Comments on Stages[8]

Note that even a simple purchase such as the one above may break down into several separate stages. A complex transaction may need to be broken down into dozens or hundreds of separate stages before it can be properly analysed.

Stages one, two, and three commence the process but create no commitment. (Rules C, D, and E have not been complied with.)

For stage four, see rule D.

At stage five, there is still no commitment, even though agreement is near. (Rules C, D, and E have still not been complied with.)

Stage six is the first offer (Rules D and E). Note that its contents are drawn from a number of previous exchanges between the parties. Now rules F to K can begin to apply.

Stage seven is a counter-offer. The buyer asks for a price reduction, plus a more onerous set of contract conditions (Rules F and G).

In Stage eight, the buyer has used a letter of intent to get work started before the contract is in place.

Stage nine is a further counter-offer (Rules F and G).

In stage ten, the buyer makes a contract by accepting the stage nine counter-offer (Rules H to L).

Stage eleven is actually unnecessary in the circumstances but may be very important commercially for payment or approval procedures of various kinds. See also paragraph 2.2.2 under The Practical Consequences.

The Practical Consequences

1 Starting Work

1.1

Never start work without a contract. Never let a contractor start without a contract unless you have discussed the position with the lawyers first and/or you can trust the contractor.

1.2

Starting immediately gets the project off to a flying start and significant schedule advantage. However, if you start work on the basis of a letter of intent or without a signed contract, you do so at your own risk. *Get management agreement* in writing.

---

8 If you want to know more about contract law, read David Wright's book *Law for Project Managers*, published by Gower ISBN 0-566-08601-8

2   Awarding Contracts

2.1

Be clear about the authority of the person negotiating and entering into a contract on behalf of the organization. As a project manager, you are unlikely to award contracts. However, if you are authorized to negotiate, mark any written documentation 'subject to contract.'

2.2

A contract requires an offer and acceptance without qualification. So make sure that there no issues subject to contract or outstanding discussions about matters, such as special conditions to be agreed at a later date.

2.2.1
Be careful how you phrase your discussions. Ask a question in clarification rather than making a statement about your requirements, which could be taken as a counter-offer.

2.2.2
Regardless of whether the purchase order is acceptance of an offer or an offer in its own right, all purchase orders for goods or work must require the supplier to acknowledge acceptance of the order. This will avoid having a situation where a contract fails to be made.

2.3

Do not forget to debrief all unsuccessful tenderers as soon as possible. Try to tell them something useful (to them), and they will thank you with a better tender next time.

2.4

Log all purchase orders and contracts into the appropriate status report system.

3   Contract Document[9]

3.1

Read the full contract. (RTFC – know it well and sleep with it under your pillow!)

3.2

Ideally you want a clean contract without attachments. Nevertheless, if there are lots of documents involved, include an 'entire agreement' clause in the terms and reference all the documents that are part of the contract. However, make sure that any of the agreed documents do not have references to other documents, which could then become part of the contract!

---

9 If you want to investigate contract terms, read David Wright's book *Using Commercial Contracts* published by Wiley, ISBN 978-1-119-15250-7.

3.3

Most contracts are a mixture of fixed elements and reimbursable elements, and the true nature of the contract (who is responsible for the risks) is determined by the words in the written contract document.

3.4

Remember if you alter a standard form of contract, you may not know what the words mean anymore, and the contract may not hang together.

3.5

You must make sure that your contract has an overall limit of liability from 'whatever cause.'

3.5.1

It is also useful to have a cap on the liquidated damages. This prevents being sued for damages at large and unlimited liability.

3.6

Try to get your own change order form incorporated into the contract.

3.7

Endeavour to get a clause limiting the overall cumulative value of variations that the client is permitted to make. Thus once the value has been exceeded, the client can only make changes with the agreement of the contractor (say a limit of 15 per cent or 25 per cent of the original contract price). Once this limit has been reached, then the rates increase.

3.8

Another option (particularly for long-term contracts) is to quote rates that are valid for, say, twelve months but are then subject to an escalation formula.

3.9

Quote rates not just for 'direct costs of technical personnel' but for all 'indirect costs of head office and back-up staff' right up to director level, on the basis that major changes need managing and are additional to normal overheads that are included in technical man hour rates.

3.10

Develop words that make sure that any change order from a subcontractor is full and complete, and no further claim can be made.

4 Contract Awarded

4.1

Check that the executed documents conform to the proposal and review any restrictions in the contract documents.

4.2

Get the records (and informal notes) of all pre-contract negotiations with the client, particularly any commitments made or restrictions imposed.

4.3

Provide the legal department with the original of the executed contractual documents, secrecy agreements, and so on for retention and distribution of confirmed copies.

4.4

Prepare, or have legal prepare, a contract synopsis covering all important aspects – in plain English.

4.5

Ensure that all key members of the team read the contract synopsis and that each team member reads, *in full,* any areas of the contract that apply to them specifically. Make sure that they fully understand the contractual position and the associated risks.

4.6

Examine the risk analysis carried out during the tendering phase and note the key areas of risk. Convey these to the key members of the team.

4.7

If necessary, perform a new risk analysis. Although the legal department will have approved the contract, ask them for their areas of concern.

4.8

Define the relationships with all other parties involved: joint venture partners, consultants, and so on. If contracts are required, ensure that they are executed quickly.

4.9

Ensure that the workweek and hours per day are established for manual and nonmanual personnel, with client approval as required or necessary.

4.10

Ensure systems exist for getting agreement from the client for overtime working if required.

4.11

All you have to do is what it says in the contract!

5 Contractual Issues

5.1

Despite knowing something about how contracts are made, don't try and be your own lawyer. Keep good relations with the legal department – talk to them.

5.2

If the contract has not been signed, make considerable efforts to get it signed. You probably will not get any money from the client until it is signed. See 1.2 under The Practical Consequences.

5.3

Ensure you give the legal department timely notice of any potentially litigious areas. For the seller, the danger areas are a verbal complaint, a written complaint, or a missed and withheld payment. For the buyer, they are a missed date or a key loss of quality.

5.4

Seek legal advice on any international trade practices, for example, boycotts or embargos.

5.5

Variation/change orders. Register your claim with the client as soon as possible and as soon as you become aware of it and, at the latest, within the time stated in the contract. Whether you pursue the claim for extra costs and time is a different matter. See Part IV, Section L Variations/Changes/Claims. You need to know the claims procedures by heart.

5.6

"The Law Courts of England are open to everyone as are the doors of the Ritz" (Lord Chief Justice Denning). Do not get into fights quoting the contract so that you end up in court. It's expensive and everyone loses.

5.7

"I will always remember my company lawyer's comment: 'What is a contract? A legal agreement enforceable upon the weaker party.'"[10]

6 Some Contractual Reminders

a. Owner/Client inputs required – record any deficiencies.
b. Data to be provided – record any deficiencies.
c. Contract administration procedures.
d. Delivery arrangements agreed – use the latest Incoterms.
e. Contact names and telecom numbers established. See Part V, Section B Coordination Procedure.
f. Free issue materials – the right quality at the right time?
g. Vendor servicemen – use common terms.
h. Inspection authority – who's?
i. Inspection by installation personnel – also use the client.
j. Match vendors' warranties to yours but exclude corrosion and fair wear and tear.
k. Make sure that any order for equipment or subcontract is not complete until the documentation (for example Operating and Maintenance Manuals) have been handed over and accepted.
l. Liquidated damages – use them as an incentivizer.
m. Maintenance period – don't forget to have a clause making adjustments for any extensions of time.
n. Remember, contracts are made amongst friends for when something goes wrong. So keep good relationships.

7 Discharge of a Contract

7.1

A contract can be terminated by a number of mechanisms listed below. Nevertheless, a project manager has fewer options, namely, to do what they said they would do in their tender and what they promised to do in the contract – option a. All the other mechanisms listed are exceptions to the rule.

a. By performance: both parties fulfil their obligations.
b. By agreement: both parties agree to terminate the contract.
c. Frustration: unable to perform due to war, riot, earthquake, and so on, or the ship has sunk.
d. Breach: the parties fail to perform their obligations.
e. Others: disablement from performing due to death or bankruptcy. Operation of an express contract term.

Regardless of the rules for discharging the contract (see Part IV, Section T Contract Completion – Close Out).

---

10 Derek Jones of Upminster Essex, in a letter to the *Times*, Friday September 26, 2014, writing about supermarkets using their power over suppliers (where they make their money).

PART IV

Project Execution

No campaign plan survives contact with the enemy.
Field Marshall Helmuth Graf von Moltke, 1800–1891

Effective Project Management: Guidance and Checklists for Engineering and Construction,
First Edition. Garth G.F. Ward.
© 2018 John Wiley & Sons Ltd. Published 2018 by John Wiley & Sons Ltd.

Section A   Project Launch

A project at rest will continue at rest unless acted upon by internal and external pressure.

Adapted from Isaac Newton

Whilst project launch is conventionally applied to launching the project execution phase it can also be applied to other phases that can be treated as projects in their own right. For example: launching a tendering and proposal process or launching a feasibility study. See Part III.

To launch a rocket you need to light the blue touch paper and release energy. Launching a project is the same. It requires the injection of energy and enthusiasm. Once you have your project in motion, it will only continue in motion if it is acted upon by internal forces – your leadership and enthusiasm. Unfortunately, for every project action, there is an equal force of inaction. Namely, there are always people who say, "You can't do that." Consequently, you must be familiar with the rules of the company, not necessarily because you are going to use them, but because you are going to interpret their flexibility in your favour. If you bend the rules, you had better be right! If you are right, it is difficult for anyone to say, "You shouldn't have done that."

The main elements involved in the project launch phase are:

- Building a team
- Understanding the scope
- Developing the execution strategy
- Developing the project plan
- Establishing the project infrastructure, facilities, and administration requirements.

Too often, not enough attention is paid to this last item. Without the right infrastructure and administration facilities, the remaining issues will only be semi-effective.

The key to completing a project ahead of schedule is to have a really efficient and effective launch phase, as indicated in Part I, Section G Achieving Success, Subsection 8. Also, an ideal situation is for the proposal team that develops the tender documents to continue as the core project team.

Under these circumstances the initial base curve of the project 'S' curve will be achieved early and the straight line portion of the 'S' curve, will be shifted to the left (see Figure IV.A.1). Provided regular progress is achieved on the straight line portion

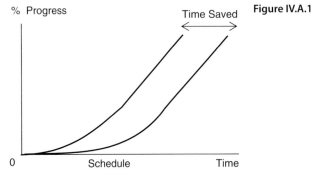

Figure IV.A.1

and the completion/close-out stage runs to plan, then the project will be completed early. The time saved at the end of the project is the same as the amount of time saved by the completion of the launch phase. See Part V, Section N, 'S' Curves.

The end of the launch phase can be defined as when the start-up curve becomes a straight line. Perversely this is not detectable until it is truly over. It requires repeatable rates of progress to demonstrate that the straight line portion of the 'S' curve, mentioned above, has been established.

John Adair's Task/Team/Individual three circles is a useful model to bear in mind during the project launch phase. The three elements need to be in balance. If you spend too much time on one element at the expense of another, then you run the risk of project failure. See Figure IV.A.2. I have added the project elements.

Figure IV.A.2 Copyright John Adair

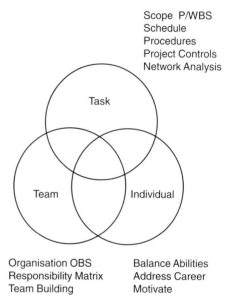

Scope P/WBS
Schedule
Procedures
Project Controls
Network Analysis

Organisation OBS
Responsibility Matrix
Team Building

Balance Abilities
Address Career
Motivate

If you do not define the task element sufficiently, then you take a piece out of the individual's motivation and similarly the team element is reduced. If the individual element is not addressed sufficiently, then a chunk will be taken out of the effectiveness of the team. Similarly, the individual may not be motivated to achieve the task. Finally, if the group is not formed into an effective team, the individuals will work to their own agendas, and the task will suffer accordingly.

The purpose of the launch phase can be summarised by the following formula:

$$\text{Performance} = \text{Capability} \times \text{Facilities} \times \text{Motivation}$$

To maximise performance, in order to shift the 'S' curve as shown above, we need to maximise the other components of the formula.

Capability: the individuals' capability is pretty much fixed. It might change minimally if some training is carried out, but in reality it is not going to change.

Facilities: this element may be only at, say, 30 to 40 per cent of its potential, but with proper attention to maximising the project infrastructure, facilities, and administration requirements (see Section C, Getting Organized), it should be possible to increase it to 60 or 70 per cent.

Motivation: this one is different. Not only can it be zero, but it can also be negative. If you have elements in the team that are negative and they don't change pretty quickly, remove them. However, the opposite is what we want; we want to maximise motivation. With good team building, this can be multiplied up by 10, 100, or even a 1000 per cent.

The following list is a quick checklist for the activities in this initial phase grouped in categories. Some of the details are developed in subsequent lists that follow.

1   Project Checks

1.1

Confirm the project job number or cost authorisation code.

1.2

Organize space for the project.

1.3

Establish key legal, intellectual property rights, and patent(s) issues.

1.4

Identify and publish the project confidentiality and security requirements.

1.5

Check any insurance needed and details of bank account(s) required.

1.6

Obtain a copy of the contract documents and/or a copy of the feasibility study. Also review the site checks carried out (see Part V, Section O).

1.7

Establish local government contacts.

1.8

Order telecommunications and information technology and computing facilities.

1.9

Establish an intranet or project website. Establish an e-mail address(es) and make courier arrangements.

1.10

Define how the various communication mechanisms are to be used. See Part VI, Section A Communications.

1.11

Book a conference room for the project duration.

1.12

Order special stationary and issue standard documentation.

1.13

Arrange for any special signs to be made.

1.14

Issue project procedures.

1.15

Obtain copies of post-project appraisals/lessons learned from recently completed similar projects. (This could save you money!)

1.16

Arrange and advise signature authority levels – project expenditure, cash advances, travel authorisation, expense reports, timesheets, and so on.

1.17

Establish the project filing system (See Part V, Section F, Filing and Archiving).

1.18

Reserve car parking – don't forget your client team requirements.

2   Project Objectives

2.1

Establish the trade-off between cost, time, and quality for the project.

2.2

Identify any conflicts with company objectives.

3   Scope Launch

3.1

Develop and confirm the product and work breakdown structures.

3.2

Check the contract documents/feasibility study/gate review comments.

3.3

Confirm the project location and/or site plan and identify the access details.

4   Team Launch

4.1

Update the project team organization charts.

4.2

Organize secretarial/clerical/admin support/mail services.

4.3

Establish signature delegation limits.

4.4

Issue the mobilization and staffing plans and set up the project roster/project directory/contact list and check work permits.

4.5

Identify which Construction Design and Management (CDM) regulations roles you fulfil (see Section Q Installation and Construction).

4.6

Hold team-building seminars. Alignment of the team's objectives is the most important aim to be achieved. See Part V, Section Q Selecting and Building the Team, subsection 2.

4.7

Arrange health and safety induction/welfare facilities/coffee machines.

4.8

Ensure that everyone is aware of the security requirements.

4.9

Discuss with the team how to maximise profit. There is the strategic decision to be made, on whether or not to make money out of changes.

5 Execution Launch

5.1

Review/validate the execution strategy/re-issue the project execution plan.

5.2

Review project procedures for the project manager's requirements before issuing them.

5.3

Identify computer hardware requirements.

6 Launch Controls

6.1

Agree on the working hours.

6.2

Order communication systems hardware.

6.3

Establish a thirty-day kick-off schedule. (See subsection 9 of this section for a list of items to be considered in the preparation of the kick-off schedule.)

6.4

Publish a project calendar.

6.5

Fix key meetings in advance.

6.6

Establish a project budget.

6.7

Establish billing/invoicing/accounting procedures.

6.8

Agree documentation format.

7 Hold Kick-Off Meeting

7.1

It is probably best to hold two separate kick-off meetings. Iron out all the issues in an internal meeting and then hold a separate (shorter) meeting with the client's team.

7.2

Despite the effectiveness of determining the project end date in the launch phase and the strong desire by an organization and individuals to demonstrate that the show is on the road, do not hold the kick-off meeting until you are ready. Resist it until you are fully prepared.

7.3

Obviously, the project team should attend and any lead discipline leaders that have already been appointed. It may be necessary to get some people released from their current assignment just to attend the meeting. It is a good idea to have your boss attend, even though you may not feel comfortable with it, which is another reason not to rush into a meeting without all the information.

7.4

Subsection 8 below provides a suggested agenda for the meeting. See Part VI, Section C for Managing and Conducting Meetings.

8 Kick-Off Meeting Agenda

8.1

Project name, number and Logo (if used)

8.1.1 Project objectives

8.2   Scope of project/work

a. Feasibility study/technology design package
b. Product breakdown structure and work breakdown structure
c. Job specification/description
d. Design specifications
e. Local codes/authorities/approvals

8.3   Scope of services

a. Other company entities involved
b. Office location and space. Facilities/furniture/reproduction
c. Project team organization/staffing. Assignment and approval of personnel. Team-building seminars
d. Productivity improvement initiatives. Value management/engineering

8.4   Coordination procedure/Administration

a. Client relationships
b. Client approvals
c. Travel/overtime/expense reports
d. Communications
e. Document procedures
f. Confidentiality
g. Project meetings
h. Progress/man-hour reporting requirements
i. Computer software programmes/systems to be used. Specialist software needed

8.5   Execution plans/Critical issues

a. Safety
b. Design
c. Spares/lubricants
d. Procurement and subcontracting
e. Inspection and expediting
f. Traffic and logistics
g. Construction and/or installation. Direct hire or subcontract
h. Prefabrication and modularisation
i. Construction camp at remote location
j. Mechanical completion definition
k. Operations and start-up.
l. Quality assurance. Project procedures

8.6   Schedule

a. Milestones
b. Kick-off schedule
c. The first 30/60/90 days

8.7 Budget/Man hours/Money

a. Cost control and estimating
b. Finance and accounting
c. Invoicing/cash flow/advance funds
d. Foreign exchange rates
e. Banking arrangements

8.8 Contract/Agreement

a. Reimbursable/non-reimbursable personnel
b. Lump sums
c. Variations/changes
d. Guarantees/liabilities/liquidated damages
e. Definition of contract completion

8.9 Site visit

8.10 Cultural issues

8.11 Immediate actions and responsibilities

9 Kick-Off Schedule

9.1

Activities to be considered in preparation of the initial kick-off schedule:

a. Receipt of the basic technology design package
b. Receipt of other client information
c. Availability of environmental data
d. Planning permission for the job site
e. Archaeology and preservation of artefacts
 f. Mobilization of the project task force
g. Project procedures finalised
h. Scope studies
 i. Development of project specifications
 j. Critical design requirements
k. Initiate basic design
 l. Identification of long-lead-time equipment and materials
m. Availability of client funds for early material commitments
n. Preparation of the preferred vendors list
o. Preparation of bulk material requisitions
p. Procurement plan and approved tenderers lists

q. Logistic studies
r. Preparation of subcontract plan
s. Construction site visit
t. Initiation of construction camp
u. Planning of temporary facilities
v. Approved trend base estimate
w. Approved project master schedule
x. Approved engineering design and construction summary schedules
y. Team building and training programmes

9.2

Examination of these activities shows that they are generally applicable to all projects. Consequently, the company that wants to make the most of the launch phase (explained at the start of this section) should have a generic kick-off schedule already available 'off the shelf'; only minor adjustments will be required.

9.3

The project controls group are the guardians of the master kick-off schedule that should be updated at the weekly project team meetings and reissued at weekly intervals.

Section B   Establishing An Office

Too often we assume that the work will be performed in the familiar environment of our company's main office. However, we may have to find space in a subsidiary office where the facilities are not as comprehensive as we are used to. In the ultimate situation you may have to rent office space and start from scratch, possibly in a foreign location.

1.1

Begin any set-up operation as early as possible. You might even take a risk and begin looking at locations and premises before the contract has been signed.

1.2

Establish a clear policy and strategy for the remote office, ensuring that everyone is clear about its terms of reference. Establish a budget for its operation.

1.3

If you are in a foreign location, employ experienced and professional consultants: in particular for taxation, accounting, and legal matters, not forgetting translation services.

1.3.1
Your home office people should have their own advisors with your home country professionals, and they should be able to use their contacts to recommend people you can use.

1.3.2
If you really are on your own, talk to your embassy or consulate and local business people to get recommendations. Go and interview half a dozen, get references, and check them out.

1.4

In theory you should register your presence with your embassy or consulate, so that they can be of assistance should there be any civil unrest. Remember your project is likely to be high profile and could be the focus of any rioting and so on. Keep in touch with them and follow their advice.

1.4.1
However, despite the above statement, do not rely on the embassy to be of any real help in time of trouble. The British Embassy wasn't, initially, to the British expatriates at the time of the first Gulf war.

1.5

In case of emergency, take guidance from the most senior company representative in the area.

1.6

Establish a detailed emergency evacuation plan. Get it approved by the right people in the home office. Be conscious of the fact that a particular job title may not make them competent in an emergency. You should be prepared to use the hidden talents of even the most junior people.

1.7

Choose your staff carefully. Your success is going to depend on them. Despite this, as the project manager, you may not be able to choose all of the expatriates. Make sure you have a right of veto on those you regard as unsuitable. For example, try not to suffer from the following communication: "Jo Brown is proposed as X job title; he has a bit of a drink problem." The response was: "So don't send him!" To which the reply was received, "I'm afraid it's too late; he is already on the aeroplane." The individual was sent back on the next available flight!

1.8

Make sure that the various office functions are managed by the respective home office functional managers. Ensure that they fully discharge their responsibilities by arranging regular audits, reviews, and management meetings in advance.

1.9

In setting up a branch office in your home country, you have an opportunity to lay out the office (subject to its physical constraints) in a manner that suits the workflow. You do not have to follow the conventions of your home office.

1.10

To reduce the possibility of fraud, adopt the mainland European convention and insist that two signatures are required on contracts and cheques and so on.

Section C   Getting Organized

We trained hard, but it seemed that every time we were beginning to form up into teams, we would be re-organized. I was to learn later in my life that we tend to meet any new situation by re-organizing and a wonderful method it can be for creating the illusion of progress while producing confusion, inefficiency and demoralisation.

Caius Petroneus, AD 66

1   Setting up the Project Infrastructure

1.1

Ensure a project, job, or cost code number is opened for the project. For internal projects, check that the correct authority level has authorised the project.

1.2

Make sure the proposal costs are cut off from the project costs. You may get stuck with proposal costs, but make sure they are recorded separately.

1.3

Organize space for the project. Try and maximise the number of people in the same location. If there will be a client in your office, it is necessary to strike a balance between integrating them into the team and task force area and preventing them from interfering at the working level.

1.3.1
Lay out the floor space in a manner that suits the workflow. This will help reduce communication problems. Further, lay it out so as to reduce the amount of time wasted moving about. Place the project files in a central location that is easily accessible to everyone.

1.4

Set up a job register system to monitor that only people working on your job can book to the job. Be ruthless about getting rid of hangers on. Update it on a regular basis.

1.4.1
Ensure that people are only assigned or removed from the project with your prior approval. If necessary, discuss assignments with the client.

1.5

Establish a weekly and/or monthly project calendar that will remain as the standard for the duration of the project. It should include public holidays, prearranged meetings and reporting cut-off dates. Update it and distribute it on a regular basis.

1.6

Fix key meetings in advance (see Section N Project Meetings). Publish a meeting schedule. Reserve a conference room for the required dates for the duration of the project.

1.7

Prepare organization charts. Issue a list of job titles and descriptions. Update them regularly throughout the project so that everyone knows who reports to whom.

1.7.1
Obtain the client's organization charts and identify interfaces with your own organization.

1.7.2
Develop detailed organization charts and manpower schedules. Identify job grades and, for a reimbursable contract, identify nonreimbursable positions. Be sensible and allow for people's holidays.

1.8

Inform the client of any proposed changes in key personnel and, if necessary, obtain their approval.

1.9

Set up any necessary employee orientation programmes.

1.10

As part of team building (Part V, Section Q subsection 2), consider using and developing a project motivational phrase to be used on all project documentation. Succinct phrases, as used by many large corporations, do make an impact. They work.

1.11

Review plans for project use of computers, software, and data processing. If you are buying software, do not buy it on the basis of a presentation from a vendor. Give them a problem or live data to use in real time.

1.11.1
This book does not address the software that is available in the marketplace. Everything is described in a manual format. Make sure that the latest software has the capability to perform these basic functions as a minimum requirement.

1.12

If you are going to make a film/video of your project, plan it now.

2   Controlling the Documents

2.1

Set up a numbering system for letters, e-mails, facsimiles, conference notes, tele-phone conversations, and so on. Ensure all communications with clients are recorded for protection down the road!

2.1.1

Each project office should maintain a correspondence log with the following information:

Serial no
Date written
Date received
Originator
Addressee
Subject file no.
Answer required – Y/N
Answer letter ref
Subject

2.1.2

Maintain a reading file with one copy of all correspondence for circulation to the project department supervisors.

2.2

Set up a master document register and establish the associated document distribution matrix. Take care – too wide a distribution can cost a fortune (remember it may be your money). Too small a distribution can cause communications to break down. See Part V, Section B, paragraph 2.17.

2.3

Maintain an action list on a daily basis and issued weekly as follows:

Item No.
Subject
Action to be taken
Date on list
Completion date
Responsible for action – company individual/client?

2.4

Set up the project filing system (see Part V, Section F). You may need an origin/destination system for letters and a subject system for other documents. Think about it carefully – you may spend an inordinate amount of time looking for papers in the future. You may need to produce letters and documents in your defence. A separate file with all documents relating to a specific problem can be very useful.

3   Responsibilities

3.1

Initiate the development of the responsibility matrix (see the project management model Figure I-B-1). The matrix can take a number of forms, from that indicated in the description of *who does what,* to forty to fifty line items of documentary processes (listed down the left-hand side). The list will show who initiates activities, who performs the actions, and who has the ultimate approval. These activities will be distributed under the headings (along the top of the chart) for the relevant corporate and project roles. It can also be used to show who reports to whom. The vertical column under a particular job title shows every contribution that is to be made by the person with that role. For each of the tasks listed on the left side of the chart, a horizontal row shows the division of responsibility for the task listed. Examination of the vertical columns will indicate where there are bottlenecks in the system.

3.1.1
Assign responsibilities and designate the appropriate authority for fulfilling the following responsibilities:

a. Purchasing commitments
b. Financial authority for cash advances
c. Signing timesheets
d. Travel requests
e. Expense reports.

   Make sure you set these at the appropriate level, relative to the project cost, so that you retain control.

3.2

If project personnel, who have been granted approval authority (for example, procurement commitment, cash advances, expense reports, or travel authorizations) no longer need this authority for any reason or are reassigned, make sure that the manager of projects sanctioning the authority is informed so that their delegated approval can be cancelled.

3.3

Set up effective, regular appraisals of personnel in the project team. If appraisals are carried out by functional managers, ensure project management has an input into the appraisals. The functional manager may have little knowledge of a person's abilities/performance on the project.

4   Procedures

4.1

Write the project procedures and, if necessary, approved by the client (if not fully detailed in the contract). If the key team members write the procedures for their areas of responsibility, review them in detail before signing off on them.

4.2

Ensure that quality assurance procedures are prepared and implemented and that regular audits are carried out. Make sure that these cover *all* aspects of the project, particularly areas of high technology.

4.3

Ensure that team members know the procedures and stick to them.

4.4

Publish a project language policy. For example, perform the project in English and then translate documents into the foreign language when they are complete. Alternately, translate the documents at the end of the project when they have to be handed over.

5   Project Execution Plan

5.1

Review lessons learnt from other projects with the team (perhaps as part of a team-building seminar). Develop, with the team, what you are going to do differently to avoid previous mistakes.

5.2

Develop the project execution plan and get the team to agree to any changes. The plan should have been developed during the tender or proposal phase (see Part III Section E, Tendering and Proposal Phase, paragraph 2.7) and should only require minor updating. Alternatively, the PEP may have originated during the feasibility stage, Part III, Section A, paragraph 4.3.

5.2.1
The summarized contents of the plan should cover the following:

a. Coordination plan
b. Resourcing and organization chart (including interfaces with others)
c. Project definition, project brief, scope of work, PBS, and WBS
d. Safety, environment, and quality plans

e. Engineering, design, use of specialist third parties
f. Procurement and (sub) contracting plans
g. Fabrication/installation/construction plans
h. Commissioning/setting to work plan
i. Systems and procedures
j. Programme – schedule
k. Budget, depending on the contract type.

6   Formalities

6.1

Establish relationships with any regulatory bodies involved in the project.

6.2

Define all approvals, which must be obtained from regulatory bodies, for example, planning permission.

6.3

Review and approve proposed permits, regulations, taxes, duties, insurance, royalties, licensing, and so on.

6.4

In a foreign location, check the requirements for visas and driving licences.

6.5

Make sure that public relations, in conjunction with the owner, handles releases to the media and reviews data, statistics, and background material.

7   Project Insurance

7.1

In conjunction with legal and insurance departments determine the insurance coverage required and authorise it.

7.1.1

Get the insurance department to check subcontractors' insurance certificates.

7.2

Ensure that the client or owner arranges any project insurance policies that they are contractually responsible for. Ensure that your own insurance expert has seen and approved the policies.

8   Some Advice

8.1

Start and maintain your project diary. I still believe in an A4 hard cover, lined paper book.

8.2

Try to persuade people to spend five or ten minutes at the end of the day, filing and keeping their desks clear. This not only assists security but also means that people have to be organized, resulting in greater efficiency.

Section D   Mobilization

The choosing of ministers is a matter of no little importance for a prince; and their worth depends on the sagacity of the prince himself.

There is one important subject I do not want to pass over, the mistake which princes can only with difficulty avoid making if they are not extremely prudent or do not choose their ministers well. I am referring to flatters, who swarm in the courts. The only way to safeguard yourself against flatters is by letting people understand that you are not offended by the truth; but if everyone can speak the truth to you then you lose respect. So a shrewd prince should adopt a middle way, choosing wise men for his government and allowing only those the freedom to speak the truth to him and then only concerning matters on which he asks their opinion and nothing else. But he should also question them thoroughly and listen to what they say; then he should make up his own mind, by himself. And his attitude towards his councils and towards each one of his advisers should be such that they will recognise that the more freely they speak out the more acceptable they will be. Apart from these, the prince should heed no one; he should put the policy agreed upon into effect straight away and he should adhere to it rigidly.

Machiavelli, Niccolo, translated by Bull, G., *The Prince*, Penguin Books 1961

1

The project manager wants all the best people who have the experience of having done the job before. However, line management must see that they are spread over all jobs to ensure continuity of the company's experience. They need to develop people's careers and be able to give them increasing responsibility.

1.1

Review the resources required with the functional managers. Remember, you do not just want the right number of warm bodies, but they must have the relevant experience. If functional managers do not give you sufficient personnel, do not hesitate to shout for help to senior management.

1.2

In the ultimate situation of conflict between the project manager and line management, senior management must either back the project manager or take them off the job. There is no middle road. But you won't make any friends.

1.3

Review the selection of the key staff with each functional manager. Check people's experience against the requirements of the project (see Part V, Section Q Selecting the Team, subsection 1 and also Part V, Section M Risk and Risk List, paragraph 1.2). Your success and the success of the project will depend on getting good people.

1.4

Try to build a team spirit, motivate, and inspire. In the final analysis, people are a resource. If they are not performing well, in spite of your best efforts, their continued poor performance will affect the whole team. So, use functional management to get rid of them.

1.5

Use functional management to resolve specific personnel and technical problems.

1.6

Integrate the secretarial requirements into the project team as a whole. Use the project manager's secretary or administrative assistant to ensure efficient and even distribution of the workload across the functions.

1.7

Establish functional links for:

a. Quality assurance
b. Project audits.

1.7.1
Make sure the functional managers perform their function within the matrix.

1.8

Do *not* mobilise personnel until they have something to do. You do not want to burn man hours for no progress.

1.9

Tell the individuals of your team that they are the 'best' – the right people for the job – and what you expect of them. Provide them with a copy of 'Completed Work' (see part V Section A, subsection 1).

1.10

Anybody who is given increased responsibilities must be told what their new duties are. If this is not done, they assume that they do their new role in the same manner as before.

1.10.1
If someone is promoted, highlight their new accountabilities. Otherwise they will do the parts of their old job that they like doing and leave the tedious and uninteresting bits to whomever takes over their old job. The consequence is that the new persons in the old job gets demotivated.

Section E   Client Relations

Good counsellors lack no clients.
> Shakespeare, Measure for Measure act I, scene I, line 35. 1604–1605

1

If you are not familiar with a new client, talk to the business development/marketing or sales person who was responsible for acquiring the client's business. Gather information and become familiar with the way that the client runs their business.

1.1

Try to remember that the client's business context and company culture (let alone their country culture, see Part V, Section C Culture) is different to yours and that, consequently, they will see the world differently. Respect the fact that, as a result, their attitudes will differ from yours.

1.2

As the project manager, work at the client interface. Establish a firm, friendly working relationship with either the client or owner. Get the manager of projects to have meetings with their counterpart. Think about how and when you can make use of your chief executive.

1.2.1
Develop informal links with other members of the client's organization. Arrange one-to-one interactions on sharing expectations, as well as big functions. Have the occasional 'off-site brief lunch' with your client project manager.

1.2.2
Encourage informal links between key members of your team and the client or owner teams. Arrange formal or working lunches as well as social gatherings.

1.3

Inform the client/owner of any proposed changes in key personnel.

1.4

Ensure that the owner or client provides all necessary information and data for the project in a timely manner.

1.5

Ensure that the client or owner approves documentation in a timely manner.

1.6

Ensure that a log is kept of the date documents are sent to the client or owner for approval so that you have a record of items that are not approved in a timely manner.

1.7

Set up any necessary procedures or plans for technology transfer.

1.8

Coordinate with the client on public relations matters. Get their permission before publishing articles about the project.

1.9

Agree with the client on the taking of photographs. See Section Q Installation and Construction, paragraph 4.3.

1.10

Think about how you are going to make use of the client. Don't forget that their success is dependent on your success.

1.11

If you have developed the right relationship with your client project manager, you can ask them for a debrief on your proposal and their evaluation of it. The timing of this is difficult. It needs to be after they feel relaxed and comfortable with you but before any controversial issues have arisen.

Section F   Scope

Poor definition of project scope is the primary cause of cost over-runs.

The U.S. Business Round Table

A project can be too big to comprehend in one chunk. Consequently, it should be broken down into smaller manageable pieces. The mechanisms used to help define the deliverables, the product to be supplied and the scope of work to be performed, are: the product breakdown structure (PBS) and the work breakdown structure (WBS). These are the fundamental building blocks of any project.

A PBS is a hierarchical breakdown of the project into its end items as a means of defining the *What* of a project. The process of developing a PBS helps to identify missing scope items and items that you hadn't thought about. The graphical presentation is a subdivision of the final project, product, or item to be produced, and it:

- Displays and defines the product to be created or produced
- Relates elements of work to each other and to the end product
- Enables responsibilities to be defined and allocated
- Forms a logical, structured, and organized base from which to combine the work to be done, the organization structure, and development of the planning and control systems.

The PBS is developed by exploding the end product into its component parts and the services required. Each package of work must be clearly distinct from all other work packages. Packages are broken down further into lower-level elements representing units of work at a level where the work is to be performed. This process of subdividing the work is continued until the project is fully defined in terms of *what* is to be done to complete the project. Although primarily oriented towards distinct self-contained end products or deliverables, software, services, and project management tasks may also be built-in. This work breakdown structure divides the work of the project into manageable parts for which responsibility can be allocated. Pamela Landy does just this in the film *The Bourne Supremacy* when she says "Now I want to break it down into boxes."

Level 1: only contains the project or end item to be produced.

Level 2: contains the major product elements or subsections of the end item. These elements can be defined by location or intended purpose.

Level 3: contains definable components or subsets of the level 2 elements.

Level 4: this is the WBS level and displays lists or units of work/activities at the level where the work is to be performed. The elements of work at this level can have costs, time, and resources (sometimes known as CTRs) allocated to them for control purposes.

Another way of looking at the P&WBS is that PBS deliverables are the nouns and the WBS level, where the work is performed, are the verbs.

Many projects can be defined in four or five levels. However, in aerospace, it may take eleven or twelve levels to get to the level where work is to be performed.

1 Scope Documents

1.1

Prepare, review, and issue the following basic project scope documents:

a. The product breakdown structure and the work breakdown structure diagrams.
b. An overall descriptive project scope – including the split of responsibilities between the contractor and client/owner. Make sure that this document is fully compatible with the PBS and WBS.
c. A technical scope document.
d. The scope of contracted services – including the split of responsibilities between the contractor and subcontractors.
e. The scope of non-reimbursable project activities.
f. The scope of management's (nonproject) contribution.

   Note: The scope must be absolutely, fully, and clearly defined. Chaos will result if it is not.

1.2

Use a negative scope list of items to clarify what is not part of the scope.

1.3

Has the client supplied all of the information required by the contract? Has any additional information or scope detail been identified, and has the client been so informed and advised of any cost or schedule consequences?

2 Changes to the Scope

2.1

Set up variation order and change order procedures (see Section L) and, if they are not already detailed in the contract, agree on them with the client. Even if they are in the contract, they will probably need clarifying.

3 Work Packaging

3.1

Use the PBS and WBS to develop work packages (see Section Q Installation and Construction, subsection 3 for its advantages).

3.2

The work packaging conceptual plan developed at the start of the project must be agreed by all disciplines: engineering design, procurement, installation, and construction.

3.3

Using work packages means that a totally engineered package can be provided to construction before work starts. This enables them to review the work and obtain the necessary resources before starting construction work.

3.4

Whilst a work package can be any size, geographical area, system, or structure for construction, it is best to limit a work package to two weeks' work during the design phase. This is particularly important for software where deliverables are less tangible.

3.5

Work packaging is invariably based on an initial split by geographical area. Then, for construction purposes, the focus is on civil engineering since they require minimal vendor information input. Construction's requirements will then move to pipe racks.

3.6

Packages must have 90 per cent definition in order to obtain effective quotations and to enable construction to procure equipment and consumables and to recruit the necessary manpower.

Section G   Estimates and Budget

When we mean to build, we first survey the plot, then draw the model; And when we see the figure of the house, then must we rate the cost of erection; Which if we find outweighs ability, what do we then but draw anew the model in fewer offices, or at least desist to build at all?

Shakespeare, Henry IV, part 2, Act 1, Scene 3

For which of you, intending to build a tower, does not first sit down and count the cost, whether he has enough to complete it? Otherwise, when he has laid a foundation and is not able to finish, all who see it begin to mock him, saying, 'this man began to build and was not able to finish.'

Luke 14:28–30

Both of the above quotations emphasize the need to define the scope. Yet it took eleven years (whilst it was being built) before the scope of the British Library was defined, Costs rose from an estimated £116m to over £550m, and it was ten years late!

Each element of the P&WBS is costed (using the best available data) and summated to determine *How Much* the total cost estimate for the project will be. At various phases throughout the project, estimates are updated and formally reviewed by management. They can be given various names as they develop, namely:

Trend base estimate	Prepared at time of project award or not later than six weeks afterwards.
Preliminary estimate	Normally used in jobs where the scope of work is very unclear at job award and a series of optimisation studies have to be carried out. Once the optimisation studies are completed, the trend base estimate is updated and renamed the preliminary estimate.
Semi-definitive estimate	This is prepared once the major items of equipment have been defined and approximate material quantities developed. This gives the estimator sufficient information to tighten the accuracy of the estimate.
Definitive estimate	Prepared when quantities of materials are accurately known and engineering design is well developed.

See also Part V, Section E Estimating and Contingency.

At the start of a project, the possibility of making an accurate estimate is, in most cases, remote. For example, you never really know how much a project will cost until you are out of the ground. Consequently, a contingency must be added. The client and contractor should reach an agreement regarding the most reasonable level for the contingency of scope of areas not yet fully defined.

As the project develops, materials and equipment are more accurately defined. The estimating department obtains this more accurate information from the cost/schedule manager who, in turn, gets their information from the trend meetings (see Section L Variations/Changes/Claims, subsection 2).

1 Establishing the Estimate(s)

1.1

Review the estimate included in the tender or proposal. How detailed is it? How accurate is it? How firm is it?

1.2

Review the estimate with the contract payment terms. How much is fixed, and how much is reimbursable?

1.3

Agree on the number of estimates to be prepared throughout the job. Try and keep the number of formal estimates to a minimum.

1.4

Schedule when each estimate is to be prepared and what input is required from each department.

1.5

Review the estimates in the following order with:

a. Your own management
b. The client or owner.

1.6

Set up a cost code and ensure items are correctly coded.

2 Trend Programme

2.1

If the proposal estimate is vague, it is vital to do a trend base estimate.

2.2

Use a trend programme to keep estimates up to date and reduce the time it takes to produce new estimates. See Section L Variation/Changes/Claims.

3   Allowances

3.1

Bulk material, equipment, and subcontract prices should, wherever possible, be based on tendered prices. However, allowances should be added to allow for:

a. Materials: cut and waste, overbuying, and the accuracy of the take-off
b. Equipment: changes to scope – scope creep
c. Subcontracts: the degree of completion of the subcontract scope and the estimated quantities to be installed. Get the construction manager to review the proposed allowances.

3.2

Labour is priced on actual costs plus allowances for future increases and overhead. However, the big issue here is productivity and how good the historical data for this aspect is.

3.3

An allowance should be considered to take into account a client's attitude. Evaluate how the client involvement in the project will influence the estimate. How bureaucratic are they? How complex or time consuming are approval procedures? A bureaucratic client will increase costs: "This is our base price, but if you wish to be involved, the price is double!"

4   The Budget

4.1

When the contract is signed or (depending on the contracting circumstances), the estimate is agreed with the client or senior management, the trend base estimate (or preliminary estimate) becomes the budget for the project.

4.1.1
Remember, the tender or proposal should have been prepared by a group of people who were motivated by wanting to win. Consequently, they may have been more interested in reducing the price than in ensuring that the estimate was correct.

4.2

You, then, have two or three days (a week at the very most) to protest at any deficiencies or errors in the estimate. After that, you will be deemed to have accepted the estimate as the budget, and any adjustments will have to be made using the contingency.

4.3

When the components of the estimate are allocated to the various discipline groups, set the budgets for the groups at 5 to 10 per cent below the contract level. This allows for in-house variations or design changes.

4.4

Plot the budget usage line (see Section K, Project Control, subsection 3, Figure IV.K.1).

Section H   Accounting

To an actuary, accountancy is exciting!

Source unknown

1   Looking after the Finances

1.1

Get the right people; see Section D, Mobilization, paragraphs 1.1 to 1.5. This will be one of the smallest groups, possibly only one or two people. With the right person, you can leave them to get on with their job, and they will give you the right information when you ask for it.

1.2

Ensure that the accounting procedures are thorough and conform to the contract and/or letter of 'intent/instruction' so that there are no delays in invoicing the owner or in making payments that are due. If you start work on a letter of intent, you do so at your own risk. With an instruction, you can get paid!

1.3

Ensure that any advance funding is invoiced without delay. Since this is the first payment on the project, it may not be a good example of how long it takes for payment.

1.3.1

Issue a normal invoice as soon as possible to discover how long it takes to be paid. Adjust the date for issuing invoices accordingly. For example, if payment is due at the end of each month, you do not issue the invoice then. The invoice should be issued on say, the twentieth of the month (assuming it takes the client ten days to approve and make payment). Performing a net present value (NPV) calculation (see Part V, Section G Financial Appraisal) demonstrates that the project can contribute additional margin to the bottom line. The invoice issue date can always be amended later as the payment process becomes more efficient.

1.4

Review and check each invoice to the owner. Monitor the cash flow, and keep the project in the black. If the project needs to borrow funds, the project will get charged the interest.

1.5

Organize the setting up of banking, foreign currency, loan, or other financial arrangements as required.

1.6

Ensure that the accounts department is aware of all agreements with the owner and third parties.

1.7

Monitor the payment terms of purchase orders for feeding into the cash-flow forecast.

1.8

Assure the protection of the owner's financial assets and cash-flow needs. Do not let them call up funds earlier than necessary.

1.9

Try to ensure that the use of soft and/or local currencies is limited to reimbursement of local costs.

1.10

Try to arrange for hard currency commitments to be invoiced and paid for in the same currency. If you spend pounds sterling, get reimbursed in pounds sterling.

1.11

If you have foreign currency loans, fix a project exchange rate and do exchange rate fluctuations as a separate report. Make sure you allocate your purchases correctly. Do not buy, say, insulation from India from a hard currency loan when you have an Indian Rupee loan!

1.12

Check whether you need to buy forward any currency. Talk to the treasury or other department responsible for these matters.

1.13

Verify that any excess cash is invested. Unfortunately, the treasury department is likely to keep any interest rather than crediting the project.

1.14

Review and approve payments, advances, fee retention, currencies, letters of credit, and so on.

1.14.1

Make sure that payments have adequate supporting documentation and that there is evidence that the work was performed. This is particularly important for progress payments.

1.14.2

Have someone who is good at dotting 'i's and crossing 't's check the letters of credit – a Belbin Completer Finisher (Part V, Section R Team Roles). Think carefully when selecting 'partial shipments not allowed.'

2   Bonds

Any bonds issued for the project are likely to have been set up in the tendering stages before the contract is signed. See Part V, Section P Surety Bonds.

2.1

Remember that bonds have a life of their own. They exist separately from the contract under the laws of the client's country, regardless of what is said in the contract. They have different life spans in different countries. They can be cashed by anyone, and they cost money every month that they are not returned to the bank that issued them. Recover them. Treat them seriously.

Section J   Planning and Scheduling

People don't like to plan it's much more fun just to do and the nice thing about just doing, is that failure comes as a complete surprise. Whereas, if you have planned, failure is preceded by a long period of despondency and worry!

John Harvey-Jones

The WBS provides the building blocks for planning the project. Planning is a group pro-cess wherein the team thrashes out the relationships between the work elements in order to determine *How* the project is to be executed. The addition of the inputs required by the work packages/activities and the outputs produced relates the work elements to each other for integration into a detailed network. With the addition of time esti-mates (using the best available data), the network can be analysed to determine *When* activities need to be performed. The sequence of activities with no spare time is the critical path.

Scheduling, on the other hand, is the mechanics of creating individual tabulations. Schedules are derived from an analysis of the network to produce lists of work/tasks for the various disciplines.

1   Getting Organized

1.1

As project manager you must be familiar and reasonably competent with the planning/scheduling system being used.

1.2

Given enough guidance, a better-than-average planner can rough out a first pass plan from an equipment list, a plot plan, and some flowsheets and come up with an end date. It won't be the date everyone wants, but it's a start.

1.3

Prepare any necessary networks. Keep the complexity to a minimum. A project must be planned to Level 3 – bar charts are just what their name says – papers to be used in bars!

2   Planning

2.1

The project manager must then sit down with the planner at the computer and key members of the project team for as long as it takes to crawl through every single activity and duration and interconnection. (Use this as a team-building exercise; see Part V, Section Q, Subsection 2 Building the Team.) Working informally together, anomalies will tumble out like leaves from a tree. Misunderstandings: ("Wow, I didn't realise that piece of equipment was that heavy"), alternatives ("We don't have to finish that before starting that"), and omissions ("We forgot about vendor data"), will be flushed out.

2.2

As indicated in Part I, Section G, paragraph 6.5, the tricky bit is how to practically amend the plan to what everyone wants, for example, pre-order critical equipment, reduce shipping time by having dedicated ships. Alternatively, bulk-order materials from hand-drawn sketches at a high risk of overages, and so on. However, always keep attacking the critical and subcritical items.

2.3

Review the proposed plan or schedule for realism. Inform senior management if the contractual end date is not possible. Never underestimate the value of a deadline, but if you give people a target that is unattainable, they will probably switch off.

2.3.1

In high-risk environments, allow for some pessimistic planning. However, it is usually possible to find alternative ways of doing things so as to be able to work around risks.

2.4

In the United States, a mutually agreed critical path network is a contractual legal requirement, intended to ensure that an owner delay or variation/change is indisputably compensated to the contractor and also to prevent the contractor's spurious claims for delay. Interestingly, a court judgement also said that float belongs to the first person to use it! See the following paragraph 2.5 below.

2.5

Make sure that people understand that they may *not* use up the float without your permission.

2.6

See Part I, Section F The Owner and Client, paragraph 2.3

3   Scheduling

3.1

Prepare a ninety-day kick-off schedule, gathering input from the engineering, procurement, and construction managers. Get the initial schedule, developed by the project controls function and reviewed by the same people, before it is reviewed jointly by the project controls manager and project manager.

3.2

Prepare a critical items list. Make sure it includes *all* critical items.

3.3

Get procurement lead times for incorporation into the schedule.

3.4

Prepare detailed lists of deliverables for all departments. Schedule dates for each deliverable. Ensure the lists are realistic. If all deliverables cannot be identified, estimate the total number. Also ensure that all the lists are comprehensive. Do not forget items such as procedures, design manuals, commissioning manuals, operating and maintenance manuals, and so on.

3.5

Issue exception reports on a regular basis. You need to know where things are *not* going right.

3.6

Carry holidays forward if you can.

Section K   Project Controls

I keep six honest serving men
(they taught me all I knew):
Their names are How and Why and When
and What and Where and Who.

"The Elephant's Child" by Rudyard Kipling

The PBS defines the **What** of the project and creates a structure from which all the other 'honest serving men' are developed. Rudyard Kipling, not being a project manager, had not yet identified two more: **What if** (the risks) and **How Much** (the estimate). These are also developed using the PBS and WBS and complete the project management model.

For control purposes, the work packages at level 4 (see Section F Scope) are assigned:

1. A description of the scope of work to be performed
2. A budget for the work derived from the estimate
3. A schedule start and finish dates representing physical accomplishment
4. The person who is responsible for the performance (quality) of the work
5. The resources required – people and materials

Project control is the principle objective of project management since it involves a process for keeping the project to the agreed plan, in order to achieve success. However, as can be seen from the work packages at level 4, it takes five plans to develop a project:

- Scope
- Cost
- Schedule
- Quality
- Resources

These five plans need to be monitored in order to identify when there are any deviations from the plans so that corrective action can be taken.

Monitoring these five plans poses the challenge of what to measure, at what level of detail, and how frequently. However, monitoring merely provides a vast quantity of data. The project manager needs to know what has deviated from the intended plan in order to make decisions about what action to take to correct matters. Control is exercised when the project manager makes decisions.[1] The quality of the decision-making process depends on two factors:

1. Pertinent and timely information
2. Experience and judgement

Remember, with regard to the information: what cannot be or is not measured cannot be controlled. It is also essential to measure what is important rather than to make important what is measured. Further, gathering the information interrupts the work process and is an imposition as far as the team members are concerned. If it doesn't provide something that is helpful to them to keep to their plans, then insufficient attention

---

1 Decision Making Project Control', paper presented by Garth G. F. Ward at Internet 1998 - 9th World Congress on Project Management.

will be given to the accuracy of the data needed. The quality of the decision-making is dependent on the project manager's style (see Part VI Section B) and their ability to harness the know-how of the project team. Thus, there are as many soft skills involved in the control process as there are hard mechanisms in the system.

1 Setting Up

1.1

Get the right people; see Section D Mobilization, paragraphs 1.1 to 1.5. Project controls is probably the most important home office group for the project manager. This group is, in essence, an extension of the project manager's function and will be doing a lot of your work for you. Choosing the right person to head up this group can go a long way to making the project successful. Make sure that you can get on with them.

1.1.1

The group encompasses both the estimating function (Section G) and the planning function (Section J). It is also responsible for processing variations, changes, and trends (Section L). In addition, they chase up reporting information and initiate the various reports (Section M). The equivalent in a multi-project context is the project office. Build the right relationships.

1.2

Ensure the cost and the planning and scheduling specialists are fully integrated into the team and take a lead role in trend meetings or programmes.

1.3

Set up a cost code and ensure that it is understood throughout the project and used uniformly.

1.4

Decide which computer programmes and digital software tools will be used for monitoring the project and get them up and running. Keep it simple.

1.5

Make sure you set up systems to monitor everything. Someone, somewhere, is going to do something stupid. You want to know before it has serious repercussions.

1.6

Do not impose your systems on subcontractors. Let them use the system they know how to operate – their own. Specify the interface data you need from them to feed into your systems.

1.7

Make sure that team or group leaders understand the monitoring and reporting system and actually use it.

1.8

Develop and approve plans for monitoring and controlling costs of:

a. Reproduction: use the project manager's secretary/admin assistant to control these costs.
b. Communication/telephones: consider using group telephones and adjust the office layout accordingly.
c. Stationery: keep special stationery to a minimum.
d. Travel and living expenses: establish hotel discounts.
e. Establish the authority level and the standard rates for courier services.
f. And so on.

2 Progress and Reporting

2.1

Ensure you get accurate weekly job hour reports.

2.2

Monitor engineering/design package development, procurement, and construction by individual progress curves and individual work package controls.

2.2.1
A good indicator of the health of the project can be seen from plotting the equipment purchase orders. See Section M Reporting, paragraph 3.4.

2.3

Decide if you will use a full earned-value system (see Part V, Section L) or if you will use a modified, simpler system. However, if you do, you will need good productivity; see paragraph 2.3.3.

2.3.1
Adopt the philosophy that an activity is only complete when it has been *received* by the next person in the work process, not when it is in someone's 'out tray.'

2.3.2
As a minimum compare the per cent of hours used with the per cent of progress made and per cent of the schedule achieved. You may also have to monitor the per cent of manpower budget spent. See Part V, Section N, 'S' Curves.

2.3.3

Monitor the productivity trend – the budget hours divided by the hours used. Greater than one is good productivity, and less than one is poor productivity. A comfortable range is 0.9 to 1.1.

2.3.4

For smaller, less-complex projects, it might be possible just to monitor the number of deliverables completed. This approach treats all deliverables as having the same difficulty, or weighting. After all, all activities need to be done. Plot the cumulative deliverables completed (planned and actual) against time. Of course, initial progress will look good because people will do the easy ones first. The difficult ones don't get any greater rewards.

2.3.5

As a productivity indicator plot, the percentage of activities completed on time against time. If the project is progressing perfectly, the line should be horizontal. As the line dips below the horizontal, you get a retrospective indication that your schedule is slipping. The steeper the line, the more the project is out of control. See also subsection 3.

2.3.6

I have always maintained that if you purchase within budget and have acceptable productivity (2.3.3 above), then, if you manage the schedule, the costs will follow.

2.4

Set up an internal change order procedure to cover for extra work generated through misunderstandings of scope, errors, and so on that will not be chargeable to the client or owner.

2.5

Train designers to refer to project management if a client representative talks to them. Get designers to ask the client representative to confirm instructions in writing before anything can proceed further.

3   Cost Progress and Control

Plot the budget/estimate usage line (see Figure IV.K.1). The budget progress/expenditure should follow the sloping line from 1 to 1. Any time the progress line goes horizontal the project is getting out of control.

4   The Critical Path

4.1

In general terms, the critical path can flip-flop between two or three critical paths, so trend the float for each of them. The critical path for process plants is invariably through piping. Rarely is it instruments and electrics.

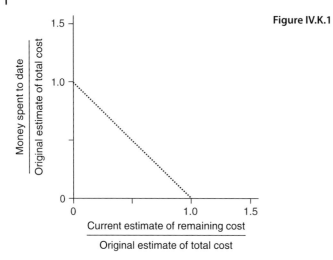

Figure IV.K.1

4.1.1
Know your critical path. One of the main critical paths for process plants is interconnecting piping layouts, material take-offs, ordering and delivery, and then pipe rack design and steel MTO, fabrication and delivery – rack/sleeper designs and construction.

4.1.2
For building work, the windows invariably seem to be critical. Get them ordered on time.

4.1.3
The piping is usually driven by the availability of materials. So, do the MTO for the straight-run materials; get them ordered and delivered early (see Section Q Installation and Construction, paragraphs 4.5 and 4.5.1).

5.1
Chase progress relentlessly.

5.2
Check costs regularly.

Section L   Variations/Changes/Claims

The loss of control of changes during design and execution is the secondary cause of project failure.

The U.S. Business Round Table

Nothing endures but change.
Heraclitus, 540–480 BC, or a modern interpretation: 'The only constant is change.'

If you throw a stone (a change) into a pond (a smoothly running project), it causes a splash (the direct impact of a change) as well as a series of ripples (the difficult-to-detect indirect impacts). If you throw lots of stones, where the ripples intermingle, disturbances with peaks larger than the individual ripples are produced (the cumulative effect of changes).

Strictly speaking, all alterations to the contract are variations when made by the client and claims (an assertion of a right under the contract) when the contractor requests compensation. Some organizations use a variety of different words to express different types of alterations; typical terminology is as follows:

- Variations are alterations to the method of working from that originally defined in the control estimate that do not impact on the direct costs. They will only adjust the allocation of costs and are internal to the organization.
- An engineering change is extra work, an alteration to the scope or time schedule or method of working requested by the client.
- A design change can be defined as a modification in scope or time schedule that is the responsibility of the contractor.
- Claims are (for changes outside the contract) when the contractor holds out a begging bowl to the client and asks for more money. It is a request for an ex gratia payment where no legal obligation exists.

Whether a client or a contractor, the only justifiable variations or changes are:

- For safety reasons
- To make the technology work
- Due to legislation or regulation
- It is essential to facilitate construction
- The change really does make a huge saving

Changes or variations to a project originate because:

- The client's business needs may change as market conditions change
- A decision is made that 95 per cent project definition is good enough
- Fallacies and optimistic statements in the project justification/feasibility study
- The requirement to accelerate the project or other operational needs
- A client's desire for preferential engineering and the difference between what the head office client contracts for and what the local client wants

Ex-gratia claims originate because of:

- Inequitable contracts created by clients using competitive tendering and awarding contracts to the lowest price, thus forcing contractors to tender artificially low to get work.

- A client's attitude that the contractor should absorb every extra cost.
- The attitude of some contractors (the civil engineering industry in particular acquired a bad reputation) to tender low and then recover money out of changes. This is exacerbated by a contractor's claims specialists and the vested interests of quantity surveyors.
- A contractor's policy about making claims in this context. Do they make claims? How do they intend to make claims, and how are the claims to be managed?

Claims can be reduced by:

- Eliminating uncertainties with precise scope definition
- Clarity concerning dates when information is required from others
- Risks should be allocated to those best able to take them or manage them
- Avoiding the use of stupidly low tenders.

Since projects go wrong half a day at a time, a process is needed in order to discover variations or changes so as to achieve good communications of changes and good project control. In addition, the regular search for trends promotes effective cost and schedule awareness in the project team. Further, client variations need to be established because they attract a *fee* that can be claimed from the client. Consequently, a system such as a trend programme must be instituted.

The objectives of a trend programme are to:

1. Keep the clients and the contractor's management informed of any variance from the project budget and schedule.
2. Keep the members of the project team informed about any changes, which could affect their work.
3. Act as an 'early warning system' so that action can be taken to solve potential problems before they become real problems.

Before a trend programme can be started, the scope, budget, schedule, type and number of personnel and desired quality of work for the project must be established. In addition, the estimate must be reviewed for any missing items.

1   Trend Base Estimate

1.1

If the work has been awarded on a fixed-price or lump-sum basis, then the estimate will be a firm estimate that can only be altered if legitimate change orders (claims) can be raised for extra work items.

1.2

Where work has been awarded on a reimbursable or 'time and materials' basis, there will be less firm information upon which to base an estimate. This estimate is, therefore, not to be regarded as a fully accurate estimate and is known as the trend base estimate.

1.3

The estimate is not a static document but is a regularly updated list of the very latest assessment of the anticipated final cost of the project. This is achieved through the trend meeting.

2 Trend Meetings

2.1

The project manager should rigorously enforce the discipline of always holding a trend meeting. Trend meetings should ideally be held once a week. They should be attended by all the senior members of the team such as:

a. Project manager
b. Project design manager
c. Lead discipline designers
d. Senior project personnel responsible for specific areas
e. Procurement manager
f. Cost and scheduling manager
g. Production/installation/construction manager or coordinator
h. Testing/commissioning manager.

It is important that members of the project team are ready to admit to having a problem at this meeting.

2.1.1
Hold the trend meeting at 11:00 a.m. to prevent it running on unnecessarily.

2.2

With so many attending, the project manager must keep firm control of the meeting. Each person, in turn, reports any deviation from the project plan. They need to keep in mind the following five factors:

Scope	Has some work been added or deducted from the original agreed scope of work?
Cost	Is the work being carried out within the cost identified in the current estimate?
Schedule	Is the work being completed on time according to the latest schedule?
Quality	Is the correct quality being maintained? Have the specifications been altered?
Personnel	Have the right number of people and resources been allocated to the project, and are the personnel of the right grade and calibre?

2.3

In listing the deviations from the plan, each participant gives as good an estimate as possible of the effect in terms of cost and schedule. If they are held up because they have not received information from another group, they briefly list the information required. If they are holding up another group, they so advise.

2.3.1

It may be the case that although a group is responsible for a delay, there is no cost and schedule delay directly involved in that group's work. However, there could be a cost and schedule impact on another group.

2.3.2

In this case, the trend meeting should not be delayed while the effect on cost and schedule is calculated. It should be noted as a trend, and the department concerned should, after the meeting, advise the cost and schedule manager and the project manager of the extent of the impact.

2.4

The cost and schedule manager gives each of these trends (each specific deviation from the plan) a number and adds a description of the trend plus the effect on cost and schedule. If something affects delivery, it affects costs.

2.5

Naturally, if a major problem occurs, the relevant group would not wait until the trend meeting to inform the project manager. The project manager will clearly want to take immediate action to avoid a project delay or an increase in cost. However, all these occurrences are announced in the trend meeting so that all participants are kept fully aware of all developments.

3   Potential Trends

3.1

Because projects are normally run on such tight timeframes, it is important to have an 'early warning system' to alert the project manager to any potential problems, which may possibly occur. Each participant at the trend meeting will not only report each trend that has actually occurred but will also report *potential* trends.

3.2

Again, these potential trends are recorded by the cost and schedule manager, and their possible effect on cost and schedule are estimated. Armed with this knowledge, the project manager can take action to find a solution before the potential problem develops into a real problem.

4   Claims for Changes

4.1

Claims for changes are sometimes viewed as an opportunity for additional income. This may be possible for significant changes, but lots of small changes are disruptive to the project and impact productivity.

4.2

An early strategic decision needs to be made as to whether the direct impact of changes are to be *costed*, or are they to be *sold* for a price, as part of a single tender negotiation.

4.3

Small claims are likely to be within the authority of the client project team to sign. Large claims may be outside their authority, and the project may not have the budget available. If the project is financed, then additional funds may be required.

4.4

It is also usually the case that the cumulative effect of many changes is severely underestimated, particularly with regard to schedule impact. See Part V, Section N 'S' Curves, subsection 2 change orders.

4.5

The project's bottom line can be improved if claims are properly recorded and procedures in the contract are followed, particularly the time for submission. Consequently, there must be one person responsible for following up change orders and seeing them resolved; see subsection 5.

4.6

A prerequisite is that all alterations are highlighted in writing within ten days after instructions are received from the client or owner (whether orally or in writing) to do any of the following:

a. Extra work not covered in the drawings or specifications.
b. Work that is different to that detailed in the drawings or specifications. 'It would help if you swapped them over ….'
c. Work in a different manner or to a different method, which is not the same as that originally planned. 'Wouldn't it be nice if … ?'
d. Work to drawings or specifications that have been changed. 'Use issue 3 not issue 2', or 'There's very little difference really.'
e. Work to documentation that is lacking in information or is incomplete and requires time to be spent developing the detail.
f. Work out of sequence or to a different method, which is not the same as that originally planned. "Could you deliver by air, not by sea?"
g. Work to one particular method when two or more alternative methods are allowed by the contract or when the contractor should be free to develop their own methods.
h. Do work when owner or client supplied information or equipment is late, in poor condition, or not suitable for the use intended.
i. Work to a different schedule, accelerate work to regain schedule, or work to a compressed schedule. Add resources or materials, work overtime, or add extra shifts.
j. Stop, disrupt, or interrupt work entirely or partially.

4.6.1

Inaction by a client must also be recorded and notified. This will cause you to employ *constructive acceleration* in order to catch up the schedule.

4.6.2

Since development of the design is the reason a reimbursable contract is awarded, it can be particularly difficult to persuade the client that design development constitutes change. Are they 'must haves' or 'nice to haves?' However, it is important to establish a change because of the associated fee.

4.7

No one should be able to approve additional work except the project manager. Starting work on a change before it has been agreed loses one's negotiating leverage.

5   Managing Claims

5.1

Use an assistant project manager as guardian of the contract and to administer the contract. They should pay for themselves. Strategies for consideration in managing claims are:

a. Train the client with small claims so that they get accustomed to the process.
b. For large claims, you need to persuade the client to 'eat the elephant one bite at a time.'
c. Use an independent work group for their execution. Use a separate cost code number.
d. Rather than disrupt work, retrofit the variation at a later stage in the project.
e. Provide an estimate of the variation, but reserve the right to adjust it for the actual cost.
f. Get paid for doing the estimate of the variation.
g. Place a percentage cap on the cumulative value of variations. See Part III, Section F Contracts, paragraphs 3.6 – 3.9 for suggested measures to be included in a contract in order to discourage the client from making variations.

5.2

Check that the liquidated damages effective date is adjusted for any schedule extensions.

5.3

Similarly, make sure that insurance policies are adjusted to reflect any changes to the scope.

5.4

There are a few fundamental rules for significant claims made against you by, say, a sub-contractor:

a. Do not try to manage the claim yourself. You will be too emotionally involved. Find the expert subcontract manager not involved with the project. Let them lead all activities, but you, the project manager, will need to be present for the sake of appearance.
b. When dealing with each element of the claim in turn, the primary question to ask is: 'Where does it say in the contract that we should pay you?'
c. Analyse the data down to the lowest level of detail. For example, say a claim was made that the weather conditions had been worse than could have been foreseen by a reasonably competent contractor. The Meterological Office evidence provided supported this. However, when the details were analysed further, the data showed that it had only rained more at night, when there was no labour on site.

6 Resist Change

6.1

In project management, 'the client is not always right.' If possible, hand the stones back to the client. Resist alterations ruthlessly, but don't be inflexible. It is, however, the job of project management to tell the client the weight of the stones and size of the splash they will make. You must advise the client of the cost and schedule consequences of their perseverance for a variation.

6.1.1
As illustrated in Part I, (Section A Project Characteristics and Phases, paragraph 3.1 Cost Impact of Decisions), after 15 per cent of the project time has elapsed and when detailed design has started, changes in scope result in schedule extension and large cost increases that you may never recover from.

Section M   Reporting

Obviously, a man's judgement cannot be better than the information on which he has based it. Give him the truth and he may still go wrong when he has the chance to be right, but give him no news or present him only with distorted and incomplete data, with ignorant, sloppy or biased reporting, with propaganda and deliberate falsehoods and you destroy his whole reasoning process.

Address to the New York State Publishers Association, August 30, 1948, by Arthur Hays Sulzberger

The challenge for reporting the **Where** of the project in the project status is:

- For what purpose is the information required?
- In what format and how should the information be presented?

See Part I Section E The Manager of Projects for what interests your boss. A generic and typical reporting structure is shown in Figure IV.M.1 below.

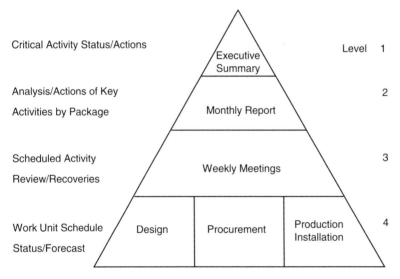

Figure IV.M.1

1   Reporting Cycle

1.1

Set up weekly, bi-weekly, or monthly (whichever is appropriate) cost and schedule reporting systems for all project groups. Make sure you receive honest and accurate reports.

1.2

A guideline for the minimum review cycle is the length of the project divided by ten or twelve. However, if the data is based on a best guess, the review cycle needs to be more frequent, for example, double this number.

1.2.1

For most projects, a week is the most appropriate reporting period. For very large extended projects, a month may be acceptable. For short duration projects, such as turn-arounds or retrofits, it may be necessary to use days or even hours for the reporting period.

1.2.2

Note: any reporting period greater than one week is asking for trouble if the schedule for the job is tight.

1.3

Set up a trend programme with weekly meetings and ensure that they work. See Section L Variations/Changes/Claims, subsection 2, Trend Meetings.

1.4

Ensure that all senior team members understand the need to monitor and control on a regular basis:

a. Scope
b. Cost
c. Schedule
d. Quality
e. Resources/personnel.

They should report any deviations from the original plan in the trend meeting.

2   Visibility

2.1

Set up a 'war room' – a conference room with plenty of flat wall space. Use this conference room as the progress reporting room, and attach as many graphical reports as possible to the wall. For example:

a. The 30/60/90 day logic diagram
b. The project bar charts
c. 'S' Curves (see Part V, Section N)
d. Resource histograms
e. Budget/estimate usage line (Figure IV.K.1)
f. Site photographs.

2.2

The visual impact of seeing the planned and actual charts superimposed or side by side can act as a motivator and even engender competition between groups.

3   Progress Reporting

3.1

The overall percentage progress will be determined from the weighting of each element at the top level. The weightings for the level 2 elements are determined by their budgeted money value. At lower levels, it does not matter what the measured units are, provided the numerator and denominator, used for calculating the percentage, are the same.

3.2

The units chosen to measure progress do not have to be the same right across the whole project. They can be mixed and matched to provide the most robust and manageable result. In the home or design office, the measurement is usually in man hours and performance measurement and earned value is explained in Part V, Section L, also see Section N, 'S' Curves.

3.3

Ensure all percentage-complete reporting is based only on deliverables completed. Never allow people to claim progress merely because they have used hours.

3.3.1

Watch that the design groups don't do the easy elements first, rather than what is required by the schedule.

3.4

Pay particular attention to the purchase of equipment and materials since this gives the best indication of the health of the project. If orders have been placed, then data and drawings can be released to design engineering. If orders have been placed, then deliveries can take place for construction needs.

3.5

Be wary[2] of construction under reporting, in the early stages, whilst you, the project manager, are still chasing up work in the home office before relocating to site. Construction has a habit of blaming the home office for holding them up for not producing the drawings and materials fast enough. They do this so that they can look good by rescuing the project later on with better-than-average performance.

3.6

Be aware[3] that all factions of the project will have different agendas and will manipulate (if they are in a position to so do) the measurement to support whatever agenda they are following. For example: contractors, especially at senior management level, will always want optimistic (that is overstated) progress in order to improve cash flow.

2,3 Advice from Vernon T. Evenson, Project Manager.

4 Progress Report

4.1

There will actually be two reports: one for the client and one for senior management. The second will be far less generic and will be significantly influenced by the individual company requirements.

4.1.1

Initiate a standard format for reporting to management. Keep it simple, realistic, and transparent.

4.2

Initiate a standard format and get agreement for formally reporting progress to the client or owner. Sometimes two monthly reports are prepared: one listing physical progress and the second a cost report – although there will be a lot of repetition to make it meaningful.

4.3

If possible, that is if the client or owner agrees, keep the textual content of the progress report to a minimum. Use charts, graphs, diagrams, and so on.

4.4

The executive report for the overall project (Level 1) will be made up from the monthly reports of the top-level components (the subprojects at Level 2) of the product breakdown structure (see Section F Scope).

4.5

Suggested contents for a monthly progress report are as follows:
   a. Cut-off date for data collection and work period covered
   b. Project manager's highlights for the period covered
   c. Problem areas and proposed actions
   d. Home office man hours and manpower histograms
   e. Construction nonmanual man hours budget expended, scheduled, and forecast
   f. Status of key subcontracts with supporting curve
   g. Major equipment and material purchases with curves
   h. Logistical highlights
   i. Summary of cost and commitments against budget and current forecast
   j. Engineering and construction progress achieved versus planned and 'S' curves
   k. Milestones achieved
   l. Status of variations and changes
   m. Forecast completion date and variations from the previous report
   n. Progress photographs
   o. And so on, but remember for what purpose the information is required

Items b to m will be summarized (one-page maximum) as the executive summary.

5   Cost Reporting

5.1

Whilst many of the other reports are focused on the individual discipline issuing the report, this report cuts across disciplines, and time spent reviewing it is essential for control of the project.

5.2

A few basic rules:

a. The allocation of cost elements must always match those used in the estimate.
b. When allocating costs, always start with the total figures and then apportion them.
c. When recording costs, always think about the *total* cost (including indirect costs), not just the direct cost of the obvious component.
d. Be very careful when comparing figures from different sources or put together in different ways – it's dangerous.

5.3

Again there will actually be two cost reports: one for the client and one for senior management. The second report will have a very defined format so that it is the same as the reports from all the other projects. Nevertheless, a cost report should consist of cost and revenue items for:

a. Major equipment
b. Bulk materials
c. Subcontract
d. Direct-hire labour
e. Nonmanual labour
f. Home office services
g. Exceptional items
h. Contingency
i. Currency fluctuation adjustment
j. Approved and pending scope changes

These items will be tabulated for each of the following headings:

i. Tender/proposal
ii. Original budget
iii. Current budget
iv. Current forecast
v. Forecast variation, over or under the current budget.

The internal report for senior management will include an item showing the forecast contribution/margin/gross margin/profit and reserves being earned.

6.1

Relentlessly chase progress.

6.2

Regularly check costs.

Section N   Project Meetings

Men are never so likely to settle a question rightly as when they discuss it freely.
        Southey's Colloquies, 1830, Thomas Babington, Lord Macaulay

1

See Part VI, Section C for Managing and Conducting Meetings. Most project meetings are a necessary part of the work process. They are, however, big man-hour consumers. Further, when people are attending meetings, they are not performing their normal productive work. Consequently, the time, frequency, and attendance at meetings should be structured so as to minimise their impact on productivity.

1.1

See Section A, subsection 7 regarding the kick-off meeting and subsection 8 for the kick-off meeting agenda.

1.2

See Part VI, Section A Communications, paragraph 4.2.2, for a morning update meeting.

1.3

See Section L, subsection 2, for trend meetings.

1.4

See Part V, Section C, paragraph 3.1, m and n, for cultural issues.

1.5

Arrange regular weekly reviews with key members of the project team and the client or owner. Encourage free and constructive interchange of information, but do not let it get out of hand.

1.6

Progress reviews are backward looking and do not control. Rather than spending time in a retrospective assessment of where the project was, we need to look forwards. This is the power of a risk-review meeting since it helps us to control by preventing change. Consequently, we should invest more time in reviewing risk. Schedule a meeting at appropriate times, depending on progress.

1.7

Arrange formal monthly reviews with key members of the project team and the client or owner. Try to get a member of senior management from both parties to attend the meetings, especially if someone is being obstructive. If possible, make use of the meeting as a business development opportunity or after-sales service.

1.8

Hold periodic performance reviews at appropriate stages (key milestones) for the purpose of learning for the future. This will maximise the capturing of knowledge before people leave the project. These stage performance reviews become a process of learning, and the post-project appraisal (see Section U subsections 2 and 3) becomes the formal process close-out.

1.9

Minutes of client meetings should be in a standard format as defined in the coordination procedure and transmitted (within three working days) to the client for their approval.

1.10

Before holding Internet meetings or video conferencing (so-called virtual meetings), make sure that people have physically met each other and that you have built the necessary relationships.

Section O   Design

Don't be buffaloed by experts and elites. Experts often possess more data than judg-
ment. Elites can become so inbred that they produce hemophiliacs who bleed to death
as soon as they are cut by the real world.

General Colin Powell

The design phase is *the* link between the owner's business development process during the feasibility stage and the production, manufacturing, fabrication, or construction
phase. More importantly, however, it is *the* link between the owner's concept for making
money and the realization of the benefits during the operational phase.

> The danger is that because of the remoteness of the home office from the physical
> activities of the project, people often start to believe that they are producing an
> end product in its own right. This is, of course, not the case - the home office is
> providing a service to the later phases of the project. It is the construction people
> who are actually producing the end product. Further, as the people who have to
> interpret the design, they need to be involved in these earlier definition stages.
> Additionally, the people in the home office need to constantly bear in mind the
> fact that small mistakes in the early paper development phases can cause a vast
> amount of work in the later physical execution stages.[4]

1   Getting Organized

1.1

Get the right people; see Section D Mobilization, paragraphs 1.1 to 1.5.

1.2

Line managers may want to introduce new systems and ways of doing things so that
the company can develop and progress. Think about it and work out how you can
cooperate.

1.3

Any contract that has been awarded after a demanding competitive tendering process
will need closer attention than normal. Make the team aware of this.

---

4 Extract from the Introduction to a Cranfield lecture note on "Project Design and Procurement" by Graham
Ritchie, 1987.

1.4

Review and approve design:

a. Objectives and plans
b. Organization
c. Staffing
d. Budgets
e. Schedules.

1.5

Clarify the need for any outside design services (other divisions, design subcontractors, consultants, design packages, and so on).

1.5.1
Subcontract design work to cheaper local offices to perform activities requiring site information, for example, steam tracing.

1.6

Arrange for any necessary design data to be obtained from the project location (site), the client, or the owner.

1.7

Arrange for any necessary visits to the project site.

2 Reviewing the Design

2.1

Think and *do* are watch words for a design group under pressure. There is a tendency to *do* and then think about it. Think and then do means: 'do it once and do it right.'

2.2

Obtain and review any process/design packages supplied by the client, owner, and or others. Ensure that they are in line with any statements made in the contract.

2.2.1
Carry out the detailed review of the basic design package with:

a. *Fabrication/installation/construction group*: to make decisions on the project break-down for subcontractors, modules and pre-assembly.
b. *Commissioning/start-up group*: to ensure the plant will start up efficiently. Agree on testing tolerances as soon as possible. See Section S, paragraph 1.2. See also Part III, Section E, paragraph 2.24.

c. *Operations personnel*: check for ease of operation and that the system or plant includes all the features they require.
d. *Maintenance personnel*: check for ease of maintenance.
e. *Safety group*: check that the plant complies with all statutory requirements.
f. *Hazard analysis group/HAZOP*: carry out a review of the process and instrument diagrams.
g. *Reliability group*: obtain their input to the design and maintenance.
h. *Environmental specialists*: check for conformance with the appropriate standards.

2.3

Ensure that client approval requirements are clearly defined and complied with.

2.4

Define the safety philosophy for the design. Design safety is 90 per cent common sense and experience, the remaining 10 per cent being special expertise. Make sure that this special expertise is used because 10 per cent can make a difference in safety matters.

2.5

Utilise functional design management to resolve technical problems. If you have doubts about a particular piece of work, get it checked out by the functional manager involved. You cannot be an expert on every aspect of design and, in any case, you do not have the time to get involved. If in doubt, get it checked. If in serious doubt, get the functional manager to confirm their approval of the design in writing.

2.6

Check the interfaces for the design sequence. Process systems get broken down to geographical areas, these are broken down again into disciplines and then to specific deliverables.

2.7

Analyse the work process and decide on the systems that are to be used for production of work.

2.8

Check contractual or financing constraints on purchasing before decisions are made about material requisitions.

2.9

If you are going to cut corners (in order to save money) talk to the quality assurance people on how you can do it without losing your quality accreditation.

3 Some Specific Design Ideas

3.1

Limit the number of vendor prints to a minimum – the essentials. Improve the turnaround of vendor prints. Time spent correctly indexing and cross-referencing vendor prints will save time later on.

3.2

Some design reminders:

a. List the equipment shown on an engineering flow diagram on the far right-hand side of the drawing. Thus the EFDs can quickly be flicked through in order to locate specific equipment.
b. Have the same scales for the piping plans and the electrical plans so that they can be overlaid to check for interferences.
c. Check that the primary voltage matches the electrical area classifications for:
    i. All vendor-supplied instruments and control valves.
    ii. Door openers and ventilation fans.
d. Make sure that the instruments listed on the engineering flow diagrams have been included prior to purchasing equipment.
e. Standardise as much as possible.
f. Design to minimise scaffolding.
g. Do not put firm dimensions on small-bore piping around equipment that has not yet been installed. Note; small-bore piping is prone to being underestimated.
h. Don't over-engineer large equipment packages.
i. On tall structures, arrange for pipe and cable trays to be on the inside of the structure and, where possible, adjacent to the stairwell. This saves scaffolding and crane erection costs on site.
j. Pay particular attention to 'undergrounds'. Look at piping crossings and ask: "How can this deep and costly excavation be avoided?"
k. Allow for fireproofing thickness on structural steel when laying out pipework.
l. Before you commit to a design change, check what stage the vendors are at. Can you make the change without disrupting them?

4 Construction Issues

4.1

Arrange for rigorous constructability reviews. Involve industrial relations in the constructability reviews to check for trade demarcations.

4.2

Optimise the extent of engineering. Only do what construction needs to build the project.

4.3

Make sure that construction defines the preferred sequence of equipment deliveries so that design and procurement can take the necessary action to achieve it.

4.4

Review the questions in Section R, if subcontracting any design work is proposed.

4.5

Allow/arrange for engineering support on site.

Section P   Procurement

> *There is hardly anything in the world that some man cannot make a little worse and sell a little cheaper and the people who consider price only are this man's lawful prey.*
>
> John Ruskin, 1850

For process plants approximately 40% of a major project is spent on the procurement of materials and equipment. It is, therefore, an extremely important area where it is possible to make significant savings in both cost and schedule, thus benefiting the overall project.[5]

Some companies have subcontract formulation under procurement and their implementation under construction. Others have all the subcontract functions within the construction department.

1   Getting Organized

1.1

Get the right people: see Section D Mobilization, paragraphs 1.1 to 1.5.

1.2

Establish functional links for:

a. Quality assurance
b. Project audit.

1.3

As already stated, if you have doubts about a particular piece of work, get it checked out by the functional management. You cannot be an expert on every aspect of the work and, in any case, you do not have the time to get involved. If in doubt, get it checked. If in serious doubt, get the functional manager to confirm their approval in writing.

1.4

Prepare procurement procedures and ensure they match contractual requirements. Ensure procedures contain all the necessary safeguards against fraud.

1.4.1
Get owner/client approval to the procedures.

1.4.2
Decide where you want to exercise your prerogative to approve certain parts of the purchasing process – for example, approving the tender lists.

---

5 From Graham Ritchie's Cranfield lecture note on 'Project Design and Procurement', 1987.

1.5

Ensure owner approval requirements are clearly defined and complied with.

1.6

Are there client requirements for the use of specific suppliers? You must register any objections you may have as soon as possible and in writing.

1.7

Check with the legal or contracts department that your standard conditions of purchase are compatible with the main contract that you have with your client.

1.8

Check financing or contractual restraints on purchasing before design decisions are made. For example, does any financing package dictate the sourcing or location of services or product suppliers?

2   Evaluating Suppliers[6]

2.1

It is the purchasing department's responsibility to build relationships with suppliers and to collect data concerning their performance. However, many of the factors considered in a supplier rating process are subjective. The following financial ratios (obtained from the suppliers' financial reports) are useful ratios that procurement can use for comparing possible suppliers:

Ratio	Use
Plant and Machinery/Sales turnover	Production efficiency
Stocks/sales turnover	Response time
Development costs/sales turnover	Development activity
Cost of materials/sales turnover	Purchasing efficiency
Cost of wages and salaries/sales turnover	Labour efficiency
Distribution costs/sales turnover	Distribution efficiency
Administrative costs/sales turnover	Administration efficiency

---

6 *The Project Manager's Guide to Purchasing, Contracting for Goods and Services* by Garth Ward. Gower 2008, covers selecting the tenderers in chapter 8.

2.1.1

Check with purchasing to see if they have or are able to make such comparisons and discuss with them what they mean in comparison to the ratios for the industry in which the company operates.

2.2

Make sure that vendors have a proven historical performance record.

3   Expediting and Inspection

3.1

Decide on the expediting policy: none/by telephone/scheduled visits. Use staff from worldwide offices. Make sure that expeditors are aware of items with zero float.

3.1.1

Decide which reports you want to receive, for example: all reports for critical items and major equipment items and only exception reports for the remainder.

3.2

Decide on the inspection policy: during fabrication at suppliers/on delivery/at site.

3.2.1

Inspect items on the critical path on a regular basis. Use site engineers to inspect at suppliers before packing. Only inspect at site those you can afford to put right if problems arise. Use staff from worldwide offices and use an inspection agency (for noncritical items) for those expensive visits. Decide on and fix client involvement.

4   Some Specific Procurement Ideas

4.1

Have strategy memos developed for the purchase of major equipment. Use them for client and management approval.

4.1.1

Fabricators of equipment packages should be selected on the basis of their ability to construct a package, rather than their expertise for the main piece of equipment.

4.1.2

Material suppliers should be prequalified on the basis of the quality of their certification paperwork systems.

4.2

Practical ideas for procurement:

a. Wherever possible reduce the procurement cycle.
b. If possible reduce the number of different vendors.
c. Include a buy-back provision in the main order for the bulks.
d. Include construction spares in the vendor's price.
e. Include vendor representatives in the price for equipment.
f. Carry out string tests for all rotating machinery at manufacturer's place of business before acceptance.
g. Get construction input for installation requirements to be incorporated into any purchase order or subcontract.
h. Get the insulation subcontractor to be responsible for doing their own material take-off prior to awarding the contract.
i. Define delivery as 'deliver onto foundations' for major equipment.
j. Rationalise the transport/shipping arrangements.

4.3

Do not buy bulks too early. Buy the minor bulks from site. However, explore the possibility of using historical data to purchase 50 per cent of the bulks early in the programme before the design is firm.

4.4

Get construction input for a consistent/common set of conditions for all vendor representatives. Include these conditions in any purchase order involving vendor representatives.

4.5

Get vendors (include a requirement in the purchase order) to communicate confirmation of the shipping of major vessels and equipment showing:

a. The purchase order number
b. The equipment number
c. Mode of transport
d. Name or number of transport vehicle
e. Estimated time of arrival.

4.6

Make sure that the subcontract terms are compatible with the main contract. See Section R Subcontracting, paragraphs 2.2 and 2.2.1.

5   Payment Terms

5.1

On the whole, suppliers' payment terms[7] should be 100 per cent upon completion of the order. Explicitly, payment is made when the goods have been delivered, including the various bits of paperwork that you need for your client. No significant payments should be made to suppliers before ownership has been transferred. A basic principle in negotiating is that you *never give something away without getting something in return*. Consequently, exceptions should be minimal; mobilization advances or payments against drawings (to keep the supplier healthy) may be made, but these must be protected by bank guarantees.

5.1.1

You might consider progress payments if the supplier or subcontractor says they would like to be reimbursed the cost of financing the order. However, make sure that the final payment is disproportionately large in order to give you leverage when you need to obtain the final documentation. If the supplier insists that they need the cash flow of regular payments, then why are you dealing with them?

---

7 *The Project Manager's Guide to Purchasing, Contracting for Goods and Services* by Garth Ward, Gower 2008, covers payment terms in chapter 10.

Section Q   Installation and Construction

> *When we build let us think we build for ever.*
>> John Ruskin. Taken from a memorial in Belgrave Square to Sir Robert Grosvenor,
>> First Marquess of Westminster

The Construction (Design and Management) Regulations 2015 [8] are meant to help the contractor:

- Improve the health and safety in the industry
- Have the right people for the right job at the right time to manage the risks on site
- Focus on effective planning and manage the risk – not the paperwork.

Everyone controlling site work has health and safety responsibilities. Checking that working conditions are healthy and safe before work begins and ensuring that the proposed work is not going to put others at risk requires planning and organization. This applies whatever the size of the site. Indicate whether you are the principal contractor. What role is the client fulfilling?

The project manager must have an understanding of the build process sequence. Within each agreed geographical area, it starts with the civil discipline of excavation for foundations, followed by the mechanical disciplines erecting steelwork and vessels and equipment. A judgement has to be made as to when there is sufficient progress to enable the critical disciplines of piping and cabling (electrical and instruments) to start. The discipline approach has then to be transferred into an area focus and finally transposed back into a systems approach to suit the requirements needed to set the facility to work. However, if resources are available, for work that does not interfere with anything else, it might as well be done – as a construction manager told me, when I asked why lamp posts were sprouting up across the site before the roads had been started.

A building process starts in the same manner: excavation and foundations, then walls and roof followed by windows, which invariably seem to be critical. When the building is water tight and secure, piping and the 'first fix' of the electrics can take place. Kitchens and bathrooms come next, followed by the finishing trades (plastering and painting), and then the 'second fix' of electrics. Finally: floor coverings, skirting boards, finishing with permanent boundary walls or fences and landscaping are done. Nevertheless, don't forget logic and common sense. The theoretical sequence will have to be modified to place underfloor heating piping in the floor screed.

The construction manager is responsible for this phase. Nevertheless, the project manager should take the initiative to move from the final construction area focus to the systems focus and the development of *start packs* needed for commissioning.

1   The Key Staff

1.1

Get the right people; see Section D Mobilization, paragraphs 1.1 to 1.5. Review and approve the installation or construction field organization and key staff. The

---

8 CDM *2015 Guidance on Regulations*. ISBN: 978071666263

choice of the installation or construction manager is the single most important choice of personnel on the project (apart from the project manager!). However, the project manager is unlikely to be able to exert much influence on the selection of the construction manager or the construction team.

1.2

For a contractor, the choice of site manager will have been determined during the tendering and proposal stage. If you are leading the proposal effort, you will have an opportunity to select the right person on the basis that they are needed as part of a team to win the contract. Make use of this opportunity. You must be able to get on with the construction manager.

1.3

If you want a successful project, you must involve the user in the earlier processes. In this case, the site manager needs to be involved in the design process. Your chances of achieving this are very slim. Site managers are out in the field until they really must be released for the next assignment. Thus, the construction phase will be started by their deputy or someone else.

1.3.1

The consequence is that a construction coordinator will be appointed to provide the construction input during the design phase. The construction coordinator must be of sufficient stature in the company to command the respect of the ultimate site management team. The decisions they make in the development of the construction plan need to be accepted and not challenged when the job becomes active in the field. Make sure that they are someone that your site manager has agreed to and has faith in. *Do not* let this person be reassigned to another project until the site manager takes over. Construction should live with their own mistakes.

1.4

The choice of subcontract manager is the most important after the choice of construction manager and needs to be chosen early.

1.5

Hold back on staffing-up until sufficient drawings, equipment, and materials are available.

1.6

Decide whether an industrial relations manager is required.

1.7

Ensure that the quality assurance department covers all installation or construction activities.

1.8

For overseas projects, maximise local hires for all nontechnical/administrative functions. Use expatriates only for management/supervisory/head of department roles.

2 Construction Planning

2.1

Is construction to be by direct hire of labour or by subcontracts? With direct hire, you achieve productivity by your own efforts.

2.2

Is construction to be cost or schedule driven?

2.3

Review and approve installation and construction budgets and reporting systems. These should be a continuation of the home office system.

2.4

If it is decided to have a construction camp at a remote location, the design, procurement, and installation of the camp may turn out to be a critical path in terms of starting work at the site. Since this will affect the overall project, it will need to be considered immediately after contract award.

2.5

Carry out a detailed review of the owner's design package to evaluate the impact on installation or construction (see Section O Design, paragraphs 2.2 and 2.2.1). Advise design of any factors that could affect their work.

2.6

Advise the design department of the proposed construction methods. Decide on the degree of preassembly and modularization, so that the design department can develop the project accordingly.

2.7

Prepare and approve a subcontract plan so that design can produce drawings accordingly. The drawings must clearly indicate the limits of a particular subcontractor's work.

2.8

Review potential industrial relations problems and plan an appropriate strategy with the corporate specialist. Review key requirements of any national industrial relations agreement or other special arrangements.

2.8.1

Develop and approve the industrial relations plan. A word of caution: it may be dangerous to assume that an expert in your home country can handle industrial relations problems in another country.

2.9

Review and approve the installation or construction plan, manpower forecasts, and so on.

2.10

Review and approve the field contract or subcontract control system including:

a. Administration
b. Reporting
c. Change orders
d. Back charges (very important)
e. Claims
f. And so on.

2.11

Check that the following have been considered in developing management plans:

a. Location: access, transport, security, permits, local taxes
b. Cultural issues: state holidays, impact of religion
c. Language: bilingual staff, availability of translators
d. Time zones: communication with home office
e. IT systems: availability locally, compatibility with home office
f. Protecting information: safes, master documents, number of copies
g. Public relations: local contacts, photographs, press releases.

2.12

Review and approve the installation or construction plan for temporary facilities (this can be as much as 33 per cent of the total cost for remote sites).

2.13

Set up any necessary craft training programmes.

2.14

Set up all necessary site procedures for:

a. Safety
b. Material handling
c. Material issuing
d. Material receiving
e. Insurance claims
f. And so on

2.15

Develop a transport policy. Do you rent or buy? What are the restrictions on importing vehicles? Do you provide minibuses for supervisors, buses for labour, and cars (depending upon their assignment conditions) for the construction manager and project manager? Nevertheless, cut down on site vehicles.

2.15.1
Control the import of vehicles. Because the paperwork was deficient, we had cars stuck in customs. After a time, customs auctioned them off to pay the storage costs they were charging! Buy locally if you can.

2.16

Develop a safety plan and a system for maintaining safety records. Evaluate different safety incentive schemes.

2.17

Develop a medical plan for remote locations.

2.18

Develop a project security plan as needed.

2.19

Develop an emergency evacuation plan.

2.20

Establish and implement manual and nonmanual personnel policies for the site and other remote locations.

2.20.1

Consult with the construction manager about looking after families, the availability of schools, and any involvement with the local community.

2.21

For remote locations, make contact with your home country embassy or consulate.

2.22

On overseas projects, ensure a system is developed for processing work permits, resident permits, security passes, and so on.

2.23

Develop mobilization and demobilization plans for manual and nonmanual personnel.

3   Work Packaging

3.1

Use a work package approach to planning and managing the work (see Section F, Scope, subsection 3.0). This enables the team to review the work, obtain the necessary resources prior to starting, and perform the work to schedule.

a. The workload can be evaluated before labour is allocated.
b. Field or site controls can be developed to suit the defined work package. The responsible supervisor or site engineer can then be monitored on the control of that individual package.
c. The status of each work package (ahead or behind schedule) can be determined more easily than evaluating the whole project.
d. Additional resources can be allocated if work begins to slip behind schedule.
e. It can be used to generate competition between similar disciplines and thus provide a positive impact on productivity.
f. Work packages provide well-defined subcontract packages, enable better control, and reduce potential claims.
g. Work packages may be modified to reflect the status of information and, if necessary, subdivided so as to maintain the flow of information to the site.

4   Construction Site Work

4.1

Do not let work start on site (building, making, and, in generic terms, any doing) before you are sure that you are ready. Also that the design team can keep feeding the hungry animal that you propose to release. Once site work starts, the construction people will start asking for information and often in an out-of-sequence order, which will add pressure to the home office work.

4.1.1
Starting work on site should be a joint agreement by the project management team. Get the project procurement manager to make a full appraisal of the material status. A firm commitment regarding delivery dates, for all outstanding materials, is needed before the decision can be made to start above-ground site erection work. This is in effect a formal 'gate review.'

4.1.2
See Part I, Section B Project Management Characteristics, paragraph 3.4.

4.1.3
As soon as the site organization is sufficient for the task, transfer all contracting activities to site.

4.2

In order to establish a really effective relationship with site management, you, as project manager, must be resident in the field full-time. You should be working site hours, at least from the start of mechanical erection. If you don't move to the site, you will not be in full control of the project.

4.2.1
The role of the project manager during construction is to ensure that construction has all the drawings and materials that they need. A key function is to manage all the interfaces and to keep the client 'off the back' of the construction manager.

4.2.2
Monitor the interaction between the site engineers and the client's engineers. This daily contact is what generates changes. Remind them that only the project managers are authorised to make changes. Consequently, make sure that the client representative(s) has the authority to give an instruction that would generate a claim.

4.2.3
The civil engineering industry is very good at using triplicate pads for site instructions: one copy to be kept by the individual receiving the instruction, one for the client giving the instruction, and one for management.

4.3

Make sure that progress photographs are taken on a regular, monthly basis from the same vantage points each time. Take additional photographs of any problems, delay factors, or additional work requested by the client. Identify the problem area on the photograph and make notes on the back as to why it was taken.

4.4

Extensive route surveys will need to be carried out for the delivery of large equipment items and any pre-assembled units. Have this done in collaboration with a specialist heavy lift/transport contractor. It will determine:

a. The limits to pre-assembled unit sizes and weights
b. The optimum route through public roads to the site
c. The optimum ship loading facilities
d. The final setting down point
e. Where roads or bridges and culverts will need reinforcement

4.4.1
Despite the route being a project effort, be aware that the police have the ultimate decision-making authority on public roads, often to the detriment of the project schedule.

4.5

Piping is usually the critical path for a process plant. Manage the balance between loading the heaviest and easiest items first, the pipe rack and piping, and manage the temptation to put off the more complex in-plant piping.

4.5.1
Get the straight-run piping materials ordered and delivered early. This enables an early build-up of the workforce and gets all the quality controls and administration in place before the difficult elements start. This obviously needs the racks in place, and it also helps to keep the lay-down areas tidy.

4.5.2
Windows are usually critical for building work. Owner clients don't have the confidence that the openings will be constructed to the accuracy detailed on the drawings and, consequently, the windows don't get ordered in time. Further, the average conventional builder doesn't necessarily have the skills required to meet the tolerances demanded by high-tech windows.

4.6

Check that all underground work (foundations, trenches, piping, and paving) will be complete before the first pre-assembled unit arrives on site.

4.7

Monitor site's use of scheduled overtime. It's expensive, and studies have shown that productivity drops off quite dramatically after five to eight weeks with regular scheduled overtime of fifty to sixty-hour weeks.[9]

---

9 NEDO report acknowledging "The Business Round Table" in the United States, concerning scheduled overtime on site. ISBN 0 11 701219X, published by HMSO in 1991.

4.8

Site cabins will be needed for foremen to view drawings and administer paperwork and so on. However, in tropical climates, do not install air conditioning. The foremen and supervisors are meant to be onsite doing a job, not hiding in a cool location!

4.9

Build relationships with the local community by hiring local resources. Sometimes the job can be done just as well and at a fraction of the cost, see w and x under 5 below.

4.10

Initiate start packs early.

4.11

Make sure that construction has a clear understanding of what constitutes project completion.

5   Some Specific Construction Ideas

a. Put hoardings/screens up around the site where it interfaces with the public. It improves safety and productivity. This is particularly pertinent for urban building sites.
b. Pre-fabricate, pre-assemble, pre-insulate, pre-dress mechanical equipment, and pre-test equipment.
c. Reduce the number of vendor representatives. Control their time and duration on site.
d. Set a maximum time for performing trade qualification tests in addition to quality checking.
e. Specify PVC sheeting instead of concrete blinding.
f. In multi-foundation areas, excavate the plot area to the elevation of 95 per cent of the foundations. Construct the foundations and then backfill.
g. Allow pockets for bolt holes.
h. Shuttering: Follow local practice and use timber off-cuts.

   Size foundations in standard increments.

i. Use a single scaffolding contractor that all subcontractors rent from. NB: scaffolding is prone to being underestimated.
j. Get the concrete reinforcing-bar subcontractor to do their own rebar schedules.
k. Award fabrication contracts (for example steelwork and pipework) to the erection contractor in order to get rid of continuous arguments about errors in fabrication.
l. Make all painting (including touch-up repairs) the responsibility of the mechanical contractors.

m. Fabricate piping spools off site.
n. Use two structural steel contractors on a design and construct basis: one for the main work and another for the miscellaneous items.
o. Construct tall structures early so as not to sterilise/isolate the surrounding area for safety reasons.
p. Schedule erection of major equipment/vessels to allow them to be placed directly onto their foundations. This minimises costly mobilization and demobilization of special/heavy lift cranes.
q. Maximise dressing of structural steelwork to the maximum extent. Fireproof off site. Erect with secondary steel/platforms in place.
r. Insulate major vessels at the vendors' works.
s. Pull cable in trenches at night and cover the trenches with sleepers during the day. Don't forget to allow for a ledge in the construction of the trench in order to allow for the timber covers.
t. When open cable trenches are necessary, install cable ducts to allow construction equipment to have access.
u. Maximise pre-shutdown work. Allow time for pre-planning; for every week of shut-down, allow one month of preplanning.
v. Where appropriate, maximise the use of local practices, particularly for building work.
w. Use two to three hundred labourers to dig a large excavation rather than buying or hiring a mechanical excavator (in China, for example).
x. On my project in Sri Lanka, we used the local elephant to string large-bore pipe (not dissimilar to tree trunks.) The pipeline was not schedule critical to the main project, and the terrain was awkward.
y. Where it is local practice, use bamboo scaffolding rather than importing steel scaffolding.

6 Establishing Authority

6.1

A word of warning: construction personnel tend to think of their function as a separate organization, and as a result, they will bypass the project manager and report to their functional manager instead.

6.2

Take control early. If necessary, find the right issue and give the construction manager a written instruction. If they ignore you, write the *facts* to your boss and wait to see how much management backing you get.

Section R   Subcontracting

The whole difference between construction and creation is exactly this: that a thing constructed can only be loved after it is constructed; but a thing created is loved before it exists.

Gilbert Keith Chesterton, Preface to Dickens's *Pickwick Papers*

Reasons for subcontracting are as follows:

- The specialist nature of the work
- A weakness in one's own organization
- In order to share or reduce risks
- Productivity of the labour force
- Competitive advantage.

Subcontracting (see Part III, Section C Issuing an Enquiry) will be the responsibility of either engineering for specialist design services or construction for site work. Both situations require the project manager's (and maybe the client's) approval, and the following questions should be addressed.

1   Questions to Ask Before Subcontracting

1.1

Are you certain that the proposed work is outside the resources of the company?

1.2

Could the company hire temporary resources to plug the deficiencies?

1.3

Are there a sufficient number of competent subcontractors capable of performing the proposed work?

1.4

Will the company have adequate experienced personnel, both technical and commercial, to supervise or manage the subcontracted work?

1.5

What additional work will the company have to perform in order to support the subcontractor?

1.6

Will the proposed construction methodology be acceptable to the subcontractor?

1.7

Can the subcontractor carry out the work in accordance with the project programme?

1.8

Will the use of a proposed subcontractor cause an industrial relations problem?

1.8.1

Will the subcontractor's management buy into your industrial relations policy?

1.9

Will the client want to get involved?

1.10

Will the client agree to the use of a proposed subcontractor, or will the client want their own choice to be used?

1.11

Is the proposed subcontract work within the budget? Can the project afford the extra costs involved in subcontracting?

1.12

The fundamental question(s) that must be answered is: what can the subcontractor do better than the company? Will the subcontractor be more effective and efficient than the company?

2   Contracting Checks

2.1

Key criteria in selecting subcontractors should be the application of innovative techniques and modularisation – not just the bottom-line cost.

2.2

Make sure that the terms and conditions for subcontractors are prepared on the basis of the master project contract between the client and your company.

2.2.1

Insert a dispute clause that binds the subcontractor to the same dispute resolution process to which you are bound.

2.3

The subcontractors must be made aware, at the time that they are asked to tender, of any overriding labour agreements at the site, together with the facilities and common services to be provided.

2.4

Ensure that the subcontractor's management undertake safety inspection audits and that the contract and the safety plan responsibilities match up.

2.5

Get the subcontractor's input to the project schedule in order to get 'buy in' to meet the schedule.

2.6

Make sure that all matters have been finalised/released by the subcontractor before closing out with the client.

3 Management Issues

3.1

See paragraph 1.4 in Section Q.

3.2

Some companies use field engineers to manage subcontracts and some companies use quantity surveyors.

3.3

Make a register of authorized signatures and authority limits for subcontractor representatives.

4 List of Some Subcontracts

a. Site survey
b. Ground conditions survey
c. Specialist ground treatment/piling
d. Bore holes
e. Construction of camp site
f. Services for camp site
g. Household and food supplies
h. Security

     i. Transport/cars/minivans/busses
     j. Air freight
     k. Shipping
     l. Courier services
     m. Supply and maintenance of site office equipment
     n. Rubbish removal
     o. Medical services
     p. Emergency evacuation
     q. Laboratory testing for materials/water quality
     r. Site clearance
     s. Temporary facilities
     t. Construction equipment/supply of tools
     u. Scaffolding
     v. Heavy lift equipment
     w. Civil work
     x. Specialist civil work/cooling towers/chimneys
     y. Special foundations
     z. Concrete reinforcing bar
   aa. Rail sidings
   bb. Storage tanks
   cc. Pipework fabrication and installation
   dd. Radiography/welder testing
   ee. Equipment installation
    ff. Structural steel supply and erection
   gg. Electrical and instrumentation
   hh. Permanent buildings
    ii. Heating, Ventilating, and Air Conditioning – Refrigeration
    jj. Connections to public services
   kk. Cathodic protection
    ll. Insulation
 mm. Painting
   nn. Landscaping and Fencing.

Section S   Commissioning and Setting To Work

> And see! she stirs!
> She starts – she moves – she seems to feel
> The thrill of life along her keel.
> 'The Building of the Ship', 1849, line 349,
> Henry Wordsworth Longfellow

On the basis of involving the user in the earlier stages, it is important to get the client's operations personnel involved in the commissioning process. Get them to start taking ownership by using them to witness tests for vendor equipment.

1

Get the right people; see Section D Mobilization, paragraphs 1.1 to 1.5.

1.1

Review and approve staffing for the commissioning representative (or team) in the home office and the team for the field.

1.1.1

Get the commissioning representative involved in the design process early on and build relationships with the client's users.

1.2

Agree the measurement tolerances for tests and proving runs as soon as possible in the design phase. As time passes, it will be much more difficult to get them approved. See similar references; Section O, paragraph 2.2.1, and Part III, Section E, paragraph 2.24.

1.3

As project manager, think about facilitating the transfer of power from the construction manager to the commissioning manager. This must happen before systems go live/hot and process fluids are involved.

1.4

Initiate commissioning start packs early in the construction phase. This involves changing the project perspective from an *area basis* to a *systems basis*.

1.5

Review and approve plans and procedures for:

a. Pre-commissioning
b. Commissioning

c. Acceptance and handover
d. Post-commissioning and start-up
e. Performance and proving tests or runs. Acceptance tests
f. Start-up spares
g. Maintenance period
h. Update health and safety requirements
i. Visits to similar operating plants
j. Visits to manufacturers of critical equipment
k. Training of client's operating personnel – involve vendors
l. And so on.

1.5.1

In an emergency, the commissioning supervisors must be able to communicate and give instructions in the local language. Consequently, institute training of your own personnel in the local language.

1.6

Have contingency plans developed to deal with issues that might arise from the following:

a. Processes without full-scale commercial experience
b. First of a kind or large scale up from previous experience
c. Prototype equipment
d. New or unknown suppliers of critical equipment.

1.7

Decide what data should be collected to improve and optimise future designs. Commissioning is a real opportunity to test the accuracy of plant design data. Over design and over specification can be eliminated; it could otherwise result in the loss of future business.

1.8

Have suppliers been advised that start-up is about to be initiated?

1.9

Keep a record of any problems involving vendors so that back charges can be raised, if necessary.

1.10

Get the client's agreement to follow-up visits in future years to find out what improvements have been made by the client since leaving the site.

1.11

A comprehensive close-out report is necessary so that accumulated 'know-how' is available for future projects. It should be included into the project historical report, (see Section U Post Project Activities, subsection 4).

1.12

This will be the last impression that the client will have of your performance, and it will be a lasting impression. The client will see at first hand any deficiencies in:

a. Organization
b. Planning
c. The time taken by the home office to respond to queries
d. Top management support when it is needed
e. The quality of the relationships at the working level

Section T   Contract Completion - Close Out

> *Now this is not the end. It is not even the beginning of the end. But is, perhaps, the end of the beginning.*
>
> Speech at the Lord Mayor's Day Luncheon, London. November 10, 1942.
>
> Sir Winston Spencer Churchill.

Project completion is often poorly defined. There is also client reluctance at taking over responsibility for the facility. Consequently, the client will try and make sure that the project is 101 per cent complete.

In the end there will, in all probability, be some minor items that it seems impossible to do. This is where the project manager will need to be involved in negotiating with the client. Do favours for the client and do a deal, swapping favours for incomplete items.

Check that all of the equipment data sheets, drawings, and so on have been reviewed and revised where necessary to incorporate all of the changes, which have been made during the commissioning phase.

Clients in building services have stated that: "The delivery of operation and maintenance manuals is the worst aspect of service because contractors leave it too late."[10]

Make sure that you leave a client who has warm feelings about you and will be receptive to future visits. This will enable you to learn about any design or project changes that have been made to improve performance

After this stage the client, having spent the maximum amount of money, hopes to start receiving a return on their investment.

1   Handover of Documentation

1.1

Original drawings and as-built drawings and any revisions.

1.2

Maintenance/mechanical catalogues and operating instructions.

1.3

Material certification, x-rays and other non-destructive testing records.

1.4

CDM health and safety file.

---

10 *Project* magazine, February 2006.

2 Handover of Equipment

2.1

Special maintenance and commissioning tools.

2.2

Commissioning and maintenance spares.

3 Clean Up

3.1

Finish painting and insulation.

3.2

Remove all temporary facilities and buildings. Ask the client if they wish to take over and use any of them. Offer some cheap modifications/painting if that will persuade the client to accept them. After all, it will be cheaper than having to remove them.

a. Office buildings
b. Stores areas
c. Training facilities
d. Lay-down areas
e. Car parks
f. Temporary fencing
g. Utilities
h. Telecommunications
i. Plant and equipment
j. Tools

3.3

Finish landscaping.

3.4

Perform a final tidy up of the site.

4 Disposal of Surplus Material

4.1

Return surpluses to suppliers under pre-negotiated buy-back agreements.

4.2

Ask the client if they wish to purchase any surpluses.

4.3

Sell surpluses on the open market.

4.4

Transfer surpluses to other projects.

4.5

Dispose of, or sell, any remaining surplus as scrap.

5 Closing Contracts

5.1

Obtain confirmation that all claims/changes/variations are completed and agreed.

5.2

Double-check that all contractual obligations have been met and that there are no unresolved matters outstanding.

5.3

Check that provisional and final acceptance certificates or a certificate of practical completion has been received from the client.

5.4

Check that all purchase orders are closed out, and the final invoices have been approved with agreement from the suppliers.

5.5

Check that all documentation required by the purchase orders has been received.

5.6

Ascertain that the supplier performance critique has been completed.

5.7

Make sure all subcontracts are closed out and the final invoices approved.

5.8

Ensure there are no outstanding matters or claims from the subcontractors.

5.9

Ascertain that the subcontract performance critique has been completed.

6 Financial Matters

6.1

Invoice for retention monies to be released: on mechanical completion, completion of performance tests, or at end of maintenance period.

6.2

Offer a retention bond in lieu of retention monies (see Part V, Section P Surety Bonds, paragraph 1.6). Don't forget to get it back!

6.3

Get agreement to the recovery of the performance bonds (see Part V, Section P Surety Bonds, subsection 2).

6.4

Close out the project bank account and any local site bank accounts.

6.5

Submit the final local tax return and obtain a tax clearance certificate.

6.6

Clear all value-added tax matters.

7 Close Out

7.1

Cancel the project insurance policies, special customs procedures, or medical arrangements.

7.2

Cancel any operating permits.

7.3

Cancel any responsible representatives.

7.4

Close out or cancel any consultancy agreements for the project.

7.5

Close out or cancel any services being provided for housing and transport.

7.6

Stop all bookings to the project, and close-out all accounts.

7.6.1

However, when closing the bookings to the project and closing out the accounts, make sure that you leave some money in the project manager's account for follow-up activities; see the next Section U paragraph 5.1.

Section U   Post Project Activities

Experience is simply the name we give our mistakes.

Oscar Wilde, 1854 – 1900.

1   Completing the Records

1.1

Issue a project souvenir. Make sure that you do not leave anyone out. Include everyone who was on the project roster.

1.1.1

If you can get the client to pay for, say, half the cost, then it will be that much easier to get management to agree to the project paying the other half.

1.2

Finalise the project video.

1.3

Write an article about the successes of the project. Professional associations and institutions are always looking for articles. Get a major supplier to write an article about the specialist features of their equipment for their trade press.

1.3.1

Find an opportunity to make a presentation about the project. Again, professional institutions might welcome a presentation for one of their meetings.

2   Post-project Appraisal – Internal Performance Review

2.1

For projects with a formal gate review procedure, the analysis of the project's performance and the lessons learned can be regarded as the last gate review activity.

2.1.1

At the end of the project, it is important to compare the final results with the original forecasts. The trend report will provide the record of changes. It is vital that detailed reasons are found for cost and programme overruns. If they are due to poor estimating or over-optimistic scheduling, the information must be fed back to the head of estimating and/or scheduling department for corrective action to be taken so that the same mistake is not made on future projects.

2.1.2

If an overrun is due to poor quality design resulting in rework during implementation, then the design group must be advised accordingly and action taken to prevent it happening on future projects. Similarly, if the overrun was due to poor morale, steps must be found to prevent a reoccurrence of the problem.

2.1.3

Statistics or metrics for estimating future projects should be collected as follows:

a. Pipework: meters per man hour
b. Electrical: meters of cable per man hour
c. Instruments: units per man hour
d. Cubic meters of concrete per item of equipment.

2.2

Areas of concern and improvements for future projects should also be looked at:

a. As a result of project location
b. As a result of the design
c. As a result of contractors or suppliers.

2.2.1

Review innovations to be used on future projects:

a. Modularisation or pre-fabrication
b. Different construction or lifting methods.

3   Project/Client Review Meeting/Lessons Learned

3.1

This is one of the few opportunities for you to get some detailed feedback on your performance as a project manager and the performance of the company. However, the focus will be on what went well, what could be done better, and how work processes could be improved.

3.1.1

For a meeting with the client to have any value, it must be on the basis of 'no holds barred', and a no-blame culture is essential.

3.1.2

Nevertheless, it will still be advisable to hold a separate meeting with the project team before having a meeting with the client for their feedback.

3.2

A list of project and contractual subject areas to be reviewed, hopefully without too much emotion, is as follows:

a. Client interface and client attitude
b. Project management and team issues
c. Corporate/functional management support
d. Quality of the proposal
e. Project execution plan
f. Planning, monitoring, and control
g. Progress reporting
h. Contract administration
i. Contract clauses wording
j. Changes
k. Quality of the estimate
l. Financial administration
m. Cost report/gross margin analysis
n. Documentation.

3.3

Emotions may run high, and it is most likely that you will need to agree that the record of the meeting will be emasculated. If you don't, people will be reluctant to talk.

3.4

The timing of these meetings can be difficult. You may have to hold a meeting before the theoretical end of the project; otherwise, many team members will have moved onto other projects. Consequentially, it is sensible to hold reviews at significant stages of the project, before key people leave the team. Alternatively, make the key people write a brief 'lessons learned' report before they (agency personnel or personnel from other organizations) are allowed to walk out of the door and leave the project. See Section N, paragraph 1.8.

3.5

To make sure of a comprehensive lessons learned process and recommendations for other future projects, the lessons learnt should be fed into the organization's 'memory bank' via topic champions. Other project managers/teams need to be made aware of the issues identified.

4 Historical Report

4.1

The historical report should contain the following:

a. The finalised cost report
b. Description and details of the project
    i. A final set of photographs
    ii. Design information
    iii. Cost and schedule details
c. Manual labour analysis
d. Details of subcontracts
e. Special features/problems/studies carried out

4.2

The information in subsections 2 and 3 above will be included in the project historical close-out report, which will list the lessons to be learned from the running of this particular job. Only by learning from past successes and failures can the company improve its overall productivity, efficiency, and profitability.

4.3

The procurement activities will probably be a separate report specifically for the procurement department. It should provide an analysis on the performance of all the vendors (provision of data, drawings, vendor servicemen, and shipping/transport activities).

4.4

You may have to start the reports well before the end of the project. You might even make it a condition that you receive someone's contribution for the historical report before you are willing to release them. This might work. You won't get it after they are on their new assignment!

5 Client Follow-up and Marketing

5.1

As indicated in the previous section, leave some money in the project manager's account for follow-up activities with the client. This will probably be contrary to the company's accepted way of working. However, its use will be the most effective client relationship building and marketing that you can do. It may win the next project.

5.2

Visit the client and find out their post-project close-out problems. If necessary, take a technical specialist with you on subsequent visits.

6 Internal Projects Benefits

6.1

After six to nine months, review the business benefits and compare them with the case submitted for project approval.

6.1.1

What happens after the project is delivered may be the primary reason for doing the project in the first place.

PART V

Specialist Topics

Section A   Completed and Inspected Work

1   Completed Work[1]

These principles of completed work are so relevant to project work that this article is reproduced with little change. The principles apply whenever a group endeavour is large enough to require division of workloads and delegation of responsibilities among *several* or *many* people. Explain the principles of completed work to your team members, and it will help you complete your project on schedule.

1.1

The supervisor must divide the work and delegate responsibilities.

1.2

The subordinate must assume these responsibilities and work out their assigned problems or duties without asking that their supervisor do part of their work.

1.3

It is far too easy, but extremely inefficient, to ask the boss what should be done rather than advising them what they should do.

1.4

Completed work is the study of an issue or problem, and the presentation of a solution, by an employee in such a form that their supervisor or department head may simply indicate approval of the *completed action*.

1.4.1

The words *completed action* are worth real emphasis. Actually the more difficult the issue is, the more tendency there is to present the issue to the supervisor in piecemeal fashion. It is the responsibility of the employee to work out the details. They should *not* consult their supervisor in the determination of these details unless necessary. Instead, if the employee cannot determine these details by themselves, they should consult other persons.

1.4.2

In far too many problem situations, the typical impulse of the inexperienced person is to ask the supervisor what to do; this recurs more often when the issue is difficult. It is accompanied by a feeling of mental frustration. It seems to be much easier to ask the supervisor what to do and appears to be so easy for them to provide the answer. This impulse must be resisted. People succumb to it only if they do not know their job.

---

1 Adapted from a work by Howard P. Mold. Source unknown.

1.4.3

It is the employee's job to *advise* supervisors what *they* ought to do; not to ask them what *the employee* ought to do. Supervisors need answers, not questions! The employee's job is to study, analyse, check, restudy, and recheck until they have come up with a single proposed action – the best one of all that they have considered. The supervisor may then approve or disapprove.

1.4.4

In most instances, completed work results in a single document prepared for the signature of the supervisor, *without accompanying comment*.

1.4.5

Except for record purposes, writing a memorandum *to a supervisor*, therefore, does *not* constitute completed work. Writing a memorandum *for a supervisor to send to someone else does*.

1.4.6

The employee's views should be placed before the supervisor in *finished form* so that they can make them their views simply by signing their name. If the proper result is reached in finished form as a solution to a problem, the supervisor will usually recognise it at once. If they need comment or explanation, they will ask for it.

1.4.7

The requirements for completed work do not put aside the possibilities of a *rough draft* in place of a highly finished form in approaching some issues. However, a rough draft must *not* be a half-baked idea! Neither must a rough draft be used as a means for shifting to the supervisor the burden of formulating the action. It must be *complete in every respect* except that it lacks the requisite number of copies and need not be neat.

1.5

Completed work requirements may result in more work for the employee, but it provides more freedom for the supervisor. This is as it should be, since it accomplishes two things:

a. The supervisor is protected from half-baked ideas, voluminous memoranda, and immature oral presentations.
b. The person who has a real idea to sell is enabled more readily to find a market.

1.6

Test the completeness of work by asking this question: as a supervisor, are you willing to sign the paper that has been prepared and stake your professional reputation on its being correct?

1.6.1

If the answer is *no*, reject it and get it worked over because it is not yet completed work.

2   Inspecting Work

Writing documents is often seen as a chore. Consequently, using modern document quality control tools such as a spell checker and a thesaurus will increase their effectiveness enormously. Nevertheless, inspection methods are still required to make sure that documents are clear, self-explanatory, unambiguous, and consistent with all other documentation used on the same project.

Inspection is aimed at project definition documentation at all levels and at all stages of their development. Guidelines are as follows:

2.1

During the project planning process, identify points where inspections will be carried out in addition to the key milestones during the project.

2.2

It is generally accepted that it is easier and more effective to use hard copy. It is easy to slip into skim reading and miss words, characters, and marks when reviewing documents on a screen.

2.3

Inspections should be limited to two hours. Any longer and the reader's attention to detail diminishes.

2.4

Carry out inspections in a prescribed series of steps.

2.5

All classes of defects in documentation should be inspected.

2.6

Produce a checklist guide for all classes of defects (see 2.7).

2.7

Each inspector is assigned a specific role in order to increase effectiveness.
For example:

a. Punctuation
b. Spelling
c. Grammar
d. Technical accuracy/correctness

e. Style, terminology
f. Formatting Layout issues dealing with illustrations, tables and so on
g. Consistency
h. Cross references and page numbers
i. Readability and does the text communicate its intended purpose.

2.8

After a couple of sessions, revitalize the inspector's interest by interchanging inspection roles/error search category.

2.9

Inspecting material should not be rushed. It should be carried out at a rate, which maximizes the discovery of errors. Over time it will be possible to determine the particular rate, which is found to give maximum error-finding ability.

2.10

Statistics on types of defects should be kept and used for analysis and reporting.

2.11

Inspections are carried out by colleagues at all levels.

Section B   Coordination Procedure

Give us the information and we will start the job.
Paraphrase of Sir Winston Churchill radio broadcast February 9, 1941. "Give us the
tools and we will finish the job."

This is one of the first project documents to be produced, if not the first for the project manager. It kick-starts the project and acts as a brief to initiate activities and the production of documents identified below. Its purpose is to regulate and standardize interactions within the project team and with the client.

Key coordination information needs to be issued as promptly as possible. Consequently, the subject matter covered by project management procedures may only be covered superficially in the coordination document. The important details will be in the procedures themselves.

1   Basic Organizing Information

1.1

Project name and number – contractor's and/or client's

1.2

Project logo and motivational phrase to be used on all project-specific documents

1.3

Statement of project objective

1.4

Home office addresses and telephone numbers

1.5

Project office addresses, that is: where the work is to be done – if it is different from the home office, together with e-mails for formal use

1.6

Contact telephone numbers, fax number

1.7

Site address, telephone number, fax number, and e-mail

1.8

Any contractor or client associate companies and their addresses and contact information.

1.9

Third-party contact details, for example, local authority

2 Coordination with the Company

2.1

Specific rules for how communication mechanisms are to be used (identified and discussed in Part VI, Section A Communications) must be defined as part of the coordination procedure, for example: see 2.2.

2.2

Rules for letters:

a. Letters are sent from project manager to project manager and restricted to a single subject per letter.
b. Letters are coded with a consecutive numbering system, for example:
   Client/Contractor (3 initials each) 001, etc.
   Contractor/Client (same 3 initials reversed) 001, etc.
   A register is maintained for each code combination. Agree on the number of copies required with the client.
   I like the client who insisted on only two digits for the letter numbers since they 'would not be writing that many letters!'
c. Transmittal letters will be coded and numbered separately and recorded in a separate register. It is sensible not to have more than one type of document per transmittal.
d. Letters between other entities will have similar codes with associated registers.
e. Any attachments and the number of copies should be recorded in the registers.
f. Letters are filed as origin – destination. Nevertheless, it is also necessary to copy letters to a subject file. See Section F Filing and Archiving, subsection 3, Master File Index.

2.3

Mail delivery and collection times and locations. List details of the schedule for a courier service (restricted to documents only) between the home office and site office.

2.4

Publish limited circulation lists on a need-to-know basis. These may need to be amended as a result of protests.

2.5

Details of the project calendar. This is a side-by-side listing of week numbers (starting at 1) matched up to dates, week ending Sunday.

2.5.1

Week ending Sunday is a psychological ploy to allow catch up time at the end of the week (if needed) without losing progress in terms of week numbers. If Sunday is the start of the week, there is no benefit to be gained. It will not be worked, and a day in the week is lost.

2.5.2

The confident project manager will not have week numbers beyond the contractual end date.

2.6

Names of project managers and key team members.

2.7

Project organization charts for both client and contractor with job titles and names. With a reimbursable contract, it may also be necessary to identify other classifications and salary grades and highlight non-reimbursable positions.

2.8

A list of key contractual documents with reference numbers and dates.

a. Contract document
b. Client drawings
c. Client procedures, standards and specifications
d. Reference to the proposal document if relevant.

2.9

Description or reference to the agreed contractual scope of work.

2.10

Reference to the agreed project schedule and budget.

2.10.1

Contract start and completion and other key project dates.

2.10.2

Any intermediate handover dates.

2.11

Contract requirements relating to formal notices, approvals, and authorisations for contractor's documents, purchase orders, and vendor's documents.

2.11.1
Description of documents requiring client approval and the number of copies of documents to be submitted for approval.

2.11.2
Submission of requests for prior approval of travel and overtime and approval notice time required.

2.12

Agreed project cut-off dates for production of various reports and submission of invoices.

2.13

Date of project manager's meeting, for example, every Thursday p.m., with location of conference room. This room should be booked on day one and reserved on the appointed day for the duration of the project

2.13.1
Date of progress meeting with client, for example, every Friday a.m., again with location of conference room, preferably as for the project manager's meeting and booked accordingly.

2.13.2
Format and the number of copies (defined in the document distribution Matrix) for the minutes of meetings and the requirement for transmittal to the client (within three working days) for their approval.

2.14

Reference to the project filing system with location (see Section F).

2.15

Definition of the software to be used for the administration functions of the project for both text and graphics.

2.16

References to the project standards and specifications to be used, together with the job title of persons responsible for the various documents. For example, if these documents have not already been referenced or published as part of the proposal document, a description as to how the company's standard procedures, standards, and specifications will be modified to meet the requirements of the contract/project.

2.17

An initial document distribution matrix will form an attachment to the coordination procedure. The matrix will identify the different types of documents and the number of copies for distribution to the various contractual parties. Scrutinise this document carefully in order to reduce unnecessary copies. Make sure you identify the documents you want to receive. See also Part IV, Section C Getting Organized, paragraph 3.1.

2.18

For convenience of reference, and in order to have all of the primary coordination information in one document, provide a list of the project manager's project procedures, as follows:

a. Procedure index
b. Project procedures
c. Project organization
d. Project roster and personnel authorizations
e. Project calendar
f. Correspondence and communications
g. Minutes of meetings/conference notes
h. Project files
i. Document distribution
j. Approval of overtime
k. Travel authorisations and expense reports.

2.19

Include any project-specific issues covering policy, procedures, and services for each department.

Section C   Cultural Issues

There is nothing more difficult to take in hand, more perilous to conduct, or more uncertain in its success, than to take the lead in the introduction of a new order to things, because the innovator has for enemies all those who have done well under the old conditions and lukewarm defenders in those who may do well under the new.
Machiavelli, Niccolo, Translated by Bull, G., *The Prince*, Penguin Books 1961.

Business is about money and people. The money aspect of this equation is fairly universal. It is the people that are different. Any project in a country foreign to one's own home country will come up against cultural barriers.

Our first problem is our ethnocentric attitude, namely, our belief in the intrinsic superiority of the nation, culture, or group to which we belong. Unfortunately, our perception of our own behaviour is that it is rational and logical. Consequently, if we wish to succeed in a different culture, we need to change our behaviour in order to adapt our attitude to the differences, since it influences the way we communicate with people.

If we reject the food, fear the religion, ignore the customs, avoid the people, we had better stay home. If you are not prepared to find out about the other person's perceptions, rules, customs, and so on – don't go abroad.

This section lists some key cultural concepts that need to be considered when doing business in a foreign environment.

1   Some Definitions of Culture:

a. It reflects attitudes and values and shapes behaviour.
b. Culture is to community what personality is to individual.
c. It is the collective programming of the mind[2]
d. It is the response to the physical and biological environment.
e. It encompasses and is composed of:
    i. Arts – Cinema, theatre, museums, architecture, sculpture
    ii. Education – formal, levels, literacy, knowledge, scientific
    iii. Language – spoken, written, official, media
    iv. Law – code, common, foreign, international
    v. Politics – nationalism, sovereignty, power, interests
    vi. Religion – philosophy, morals, beliefs, taboos, rituals, holidays
    vii. Social organization – structure, status system, authority
    viii. Symbols – dress, headgear, flags, uniforms, badges, honours
    ix. Technology – invention, transport, communications, energy
    x. Values and attitudes – time, work, wealth, changes, risk

1.1

For some reason, we accept the different meaning of words but not of behaviour. We expect people to behave the way we perceive them from the media.

---

2 G. Hofstede, 1984.

2   A Seminal Grouping of Cultures[3]

Monochronic People	Polychronic People
(For example North Americans)	(For example Sri Lankans and Indians)
They do one thing at a time. They concentrate on the job.	They do many things at once. They are highly distractible and subject to interruptions.
They take time commitments (deadlines and schedules) seriously.	They consider time commitments an objective to be achieved, if possible.
They emphasize promptness.	They base promptness on the relationship.
They are low context and need information.	They are high context and already have information.
They adhere religiously to plans	They change plans often and easily.
They are concerned about not disturbing others; they follow the rules of privacy and consideration.	They are more concerned with those who are closely related (family, friends, close business associates) than with privacy.
They show great respect for private property; they seldom borrow or lend.	They borrow and lend things often and easily.
They are committed to the job.	They are committed to people and human relationships.
They are accustomed to short-term relationships.	They have a strong tendency to build lifetime relationships.

2.1

From the above it can be seen that some cultures are very *results-oriented* and the others are dominated by *relationships*.

2.2

North Europeans and North Americans think of themselves as the norm in cultural terms, but their results culture is in the minority, representing only one-fifth of the world. The remaining four-fifths is a relationship culture.

3   Some Cultural Issues to be Aware of

3.1

These generalisations have been shown to be useful:

a. Our first cultural experience in going to another country is at immigration and customs. If there is a mistake in the documentation or if it is not all there, our first reaction is that they are difficult people. However, it was our mistake in getting the paperwork wrong.

---

3 From Edward T Hall, *Understanding Cultural Differences*. Intercultural Press.

b. Northern Europeans like important matters to be in writing. Southern Europeans have a much more oral culture. Their first reaction to receiving written communications is not 'what does this mean' but 'why has this been written down?'

c. The British tend to challenge the boss and go in different directions, and as a result do not achieve as much as they should. They adopt a negative 'devil's advocate' attitude.

d. U.S. people accept the supervisor as the boss and, as a result, go in the same direction and achieve more. They adopt a positive 'can do' attitude.

e. The Japanese have a greater desire for strong authority than Westerners (British).

f. "Norwegian contractors work in a very functional way[4]: they don't need to interpret drawings or add value solving on-site problems. They just do what they are told, sometimes rather blindly."

g. In Arab countries, greetings are lengthy (up to fifteen minutes). There are long handshakes, more body contact, and less personal space. There is great emphasis on maintaining harmony and confrontation avoidance. It is good manners to let the person on your right-hand side go through a door before you. Showing the soles of the feet/shoes in Arab countries is insulting – the Turks, in particular, will be deeply offended.

h. The Japanese build relationships over four to five meetings. They spend time planning and obtaining a consensus. However, they can't cope with change.

i. The British spend little time planning and spend a lot of time firefighting. They are good at coping with change.

j. Americans like to have scheduled plans and transact. The British don't like to have instructions for achieving their objectives; they like to understand the strategic picture and develop relationships.

k. In the hierarchical societies of East Asia it is an error to tell the boss that they are wrong.

l. In African and Far Eastern cultures, age is respected for its wisdom and experience. The Arabs respect grey hair for the same reasons. Consequently, younger members in a meeting will not express their views because of their respect for the views expressed by older people.

m. Different cultures also use meetings in different ways:
   i. The British and Dutch use meetings to resolve issues, and everyone contributes and debates the subject matter.
   ii. In Germany the meeting is used for experts to exchange information.
   iii. In France they tend to use meetings for the manager to announce decisions that have been discussed and decided by management outside the meeting. It is not a discussion forum.
   iv. In Portugal there will be several discursive meetings before coming to a result.
   v. In Arab cultures the discussion is much less open. There is a reluctance to acknowledge the full extent of a problem. The information needs to be conveyed in a very delicate manner by a person of appropriate seniority.
   vi. You cannot have the concept of 'any other business' at the end of a meeting agenda in Arab, Chinese, or Japanese cultures.

---

4 Jason Hesse, 'Keeping on Track', *Project*, Spring 2016.

n. Tabling the agenda in British English means putting the agenda on the table in order to raise the subjects for discussion. In American English it means almost the opposite – to put it to one side and discuss it later.

o. U.S. people take their jackets off in order to be serious about a discussion. The Germans; however, would view this behaviour as not being serious; they keep their jackets on when being serious.

p. In the UK and U.S.A, relaxing at a meeting with one's arms folded would be interpreted as arrogance by the Finns and would be taken as a sign of disrespect in Fiji.

q. Tapping the nose in the UK would communicate that the subject is confidential, whereas in Italy it would represent a warning.

r. Placing the right thumb and index finger together to form a nought, means zero in France, OK in Britain and North America, and in Japan it represents money, but in Germany and Brazil, it is obscene.

s. The Dutch say what they mean. They have a single understanding to the meaning of an English word. If you are not used to them, they can be seen as rude to the British. The British/English do not necessarily say what they mean. Germans speak their mind and get to the point quickly.

t. Yes and no are used differently in Western and Eastern cultures.
    i.   Sri Lankans and Indians will nod their heads in a Western 'No' fashion and still mean 'Yes.'
    ii. Egyptians will say 'Yes' in a desire to please and because it is what you want to hear.
    iii. Eastern cultures will say 'Yes', agreeing to a negative statement to which a Westerner would say 'No.'
    iv. In Greece a nod can also mean 'No'.

3.2

As distance diminishes, the cultural groupings also get smaller, for example, nationalism, regional groupings, and local town rivalry. Consequently, cultural behaviours may be modified and many less-apparent cultural idiosyncrasies can become problems.

4   Management Style

The culture of your project will be dependent on your management style. Where are you on the Tannenbaum and Schmidt continuum? (See Part VI, Section B Leadership and Motivation, subsection 1.) Will you adopt an autocratic style or empower your project team to operate within limits that you set? If you have a small project team, your view will be the philosophy of the team. However, unless you impose a very strong manner, a large team will create their own culture/philosophy.

Section D   Documentation

"The historian, essentially, wants more documents than he can really use."
'Prefaces,' The Aspern Papers. *Henry James.*

A major problem for the project manager is the quantity of documents on a project. Which ones should you approve before they are issued? Which ones should you review, and for which ones do you need for information? Then there are the documents that the project manager generates themselves.

There may be 120 to 150 different types of documents produced by the project. In addition, on a major technological project, there could be in excess of 50,000 drawings. Further, each of these can be issued several times.

In today's world, much of the documentation will be computerized. Nevertheless, people do like to use hard copies. This hard copy document then acquires many notes and other annotations and, consequently, there will be a reluctance to transfer all of this information onto the latest issue of the document concerned. Making sure that people are not using a document that is out of date and ensuring that everyone is using the latest issue, requires thorough policing by, say, the project control manager's team or the project office.

Document management software is useful in providing version control, ensuring that only the most recent approved version is used. It also provides increased document security by controlling individual's access. In addition, it provides a 'paper' trail of who is accessing what and when. Further, it can facilitate archiving.

To assist with document security, try and persuade people to spend five or ten minutes filing at the end of the day and keeping their desks clear.

1   Contractor's Own Documents and Drawings

1.1   General

a. Coordination procedure (see Section B).
b. Document distribution matrix (see Section B, paragraph 2.17).
c. Filing index (see Section F).
d. Project instructions.
e. Project roster – a list of people working on the project whether full time or part time and when they have their holidays booked.
f. Procedures. Tailor these to suit your project management philosophy. Make sure you amend them to include your specific review or approval points. An example is reviewing or approving tender lists.
g. It is useful to have a list of the company's preprinted standard forms that are required to be used. Attach this list to, say, the procedure index for ease of reference.
h. Progress reports: Many company reports are likely to have a standard format. The project manager needs to identify those reports that they wish to receive from others, in order to maintain control. They also need to identify the reports that they themselves will generate or that will be generated on their behalf.

There are two main categories: internal to the contractor and external for the client/financial organizations, such as banks, finance agencies/government organizations, offshore supplies office, Department of Energy. Whilst reports are generally issued monthly, vary this to suit the circumstances.

i. Minutes of meetings. (See Part IV, Section N Project Meetings.)
j. Quality assurance reports.

1.2   Project Controls

a. Estimate – as sold. This should not change unless there are agreed variations.
b. Cost report, Commitment and expenditure, Budget and forecast. This will have two versions – internal and external.
c. Master programme, covering the basic technology or process, design, and engineering, drawing preparation, procurement and manufacturing/erection/construction.
d. Detailed schedules, 30-, 60- or 90-day look-ahead programme.
e. Change order log, variations. Contract change register.
f. Productivity analysis of the various design groups.
g. 'S' curves showing overall project progress against planned, man hours expended and forecasts, also individual group curves with the same data. These 'S' curves are a key control tool for the project manager. See Section N.
h. Manpower histograms.
i. Critical path analysis.

1.3   Accounting

a. Loan finance report
b. Foreign exchange status
c. Vendors payments schedule
d. Subcontractors payment schedules
e. Invoices to client and others
f. Payments due.

1.4   Design/Engineering

a. Design report
b. Requisitions, requisition index
c. Data sheets
d. Specifications
e. Equipment schedule or list – 'tagged' items
f. Material take off for bulk items
g. Dimensional and weight control
h. Drawing index
i. Lists of specific specialist items, depending on the technology. For example: instrument lists, steelwork schedule, or software modules
j. Estimated drawing progress, drawings issued
k. Vendor drawing and document index
l. 'By exception' reports late-start and late-finish activities, late construction release and 'hold' registers.

1.5 Procurement

a. Purchase order status report
b. Exception reports on order placement
c. Material progress schedule
d. Material progress summary – problems
e. Procurement progress
f. Expediting and inspection planned visits
g. Expediting reports – review these for items on the critical path and subcritical paths
h. Inspection
i. Orders shipped without clearance.

1.6 Construction

a. Construction manager's report
b. Subcontract status report
c. Direct labour returns
d. Key activities report
e. Material status: overages/shortages/damaged – critical items
f. Construction staff list
g. Direct labour returns.

2 Vendor Drawings and Documents:

a. Drawings and documents for review.
   Designers and engineers tend to ask for too many drawings and documents for review. Remember every drawing requested for review has to be chased up, logged, reviewed, transmitted, followed-up, and so on. Challenge them to exercise restraint without compromising the integrity of their work. This can save many man hours.
b. Documents for construction release.
c. As-built drawings and other documents for operating and maintenance manuals.

3

Remember, all documents will be issued several times and thus are a major problem, and the most important concern to be addressed is making sure that all members of the team are using the same and latest issue.

4

See also Section F Filing & Archiving, subsection 3, Master File Index.

Section E   Estimating and Contingency

An estimate should comprise of the following:

- A description of the project scope
- A list of information on which the estimate is based
- The proposed contract strategy
- The assumed installation techniques
- The project schedule milestones
- The currency exchange rates used and estimated inflation
- The assessment of project risks and contingency
- The base date of the estimate and its estimated accuracy.

The principles of estimating are the same for all disciplines, and it involves a number of distinctly different components:

- A database
- An algorithm
- The effect of people's attitudes
- Definition of scope
- Definition of method of execution
- Identification of the risks.

The estimating approach depends on:

- The status of the estimate basis (see 2c below)
- The end use of the estimate:
    - funding of the project
    - preparation of tender documents
    - tendering for project
    - finalizing project funds
- The need for details
- Previous historical data
- The tools available
- The time and budget available for preparing the estimate.

This last item is perhaps the overriding determinant. If a senior executive asks for an 'off the cuff' number, it could be anything from +/−50 per cent to +/−100 per cent, depending on our experience and ability to interpret the crude project description. The problem is that the first number quoted is always remembered. Consequently, one should quote a *range*. However, if we are allowed to spend £500,000, and have three months, we could probably give a figure accurate to within +/− 3 per cent.

1   Types of Estimate

There is a hierarchy to estimates (see Figure V.E.1), and unfortunately every company or industry sector has their own definition. The following are reasonably generic definitions.

1.1

A *conceptual order of magnitude estimate* is based on past costs or estimates of costs of similar plants using cost versus capacity curves and adjusted for scope, size and escalation. Other factors which may have to be taken into account are plant location and time of construction.

Accuracy varies from −30 to +40/50 per cent.

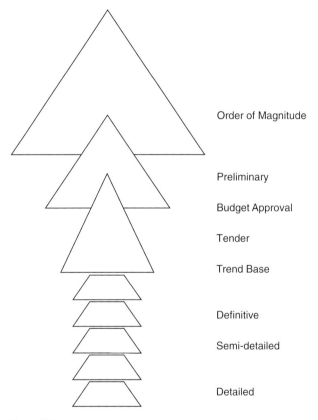

Order of Magnitude

Preliminary

Budget Approval

Tender

Trend Base

Definitive

Semi-detailed

Detailed

Figure V.E.1

1.2

A *preliminary feasibility estimate* is based on costings using in-house data or vendor quotes for items of equipment with preliminary specifications and/or data sheets and factoring the cost of bulk items (see Part I, Section F The Owner and Client, Paragraph 2.1).

Accuracy +/−25 per cent.

1.3

A *detailed tender/appropriation estimate* is compiled from data sheets for equipment, P&IDs, piping specifications, preliminary plant layout and elevation drawings, construction studies, detailed 'take offs' for all bulk materials. They are all priced from vendor's quotations and from reliable in-house data.

In order to be meaningful, that is to be successful, the overall accuracy needs to be +/−5 per cent.

1.4

A *definitive estimate* is produced during a contract at or near completion of engineering. It is put together when equipment orders have been placed and quotes obtained from vendors for bulk materials based on preliminary layouts. At this stage the construction plan and work packages representing approximately 30 per cent of the estimate will be sufficiently well defined for them to be estimated within +/−10 per cent, giving an overall estimate accuracy of +/−3 per cent.

1.5

Why knowing accuracy is important.

a. Sensitivity to economics
b. Cash flow/budgeting
c. Determine which estimating method/approach to use
d. Know where to improve methods
e. Helps establish contingency levels
f. Develop confidence/management support.

2   Estimate Planning Sequence

Estimating is a project in its own right and a team effort between project management, design, procurement, and installation. It involves:

a. Kick-off meeting.
b. Prepare estimate plan:
   i. Define responsibilities
   ii. Define work breakdown structure (and code of accounts)
   iii. Develop contracting plan
   iv. Establish estimate schedule
   v. Establish estimate reviews.
c. Prepare estimate basis:
   i. Define quantity details
   ii. Define productivity factors
   iii. Define material unit rates
   iv. Define labour and subcontract rates
   v. Define indirect costs
   vi. Define exchange rates.
d. Reviews:
   i. Project review
   ii. Project management review
   iii. Company management review.
e. Presentation to management or client.

3 The Estimating Process

3.1

The following examples are to give guidance on the concepts involved with different methods of estimating for different types of estimate.

3.2

For a conceptual order of magnitude estimate, an exponential method can be used:[5]

$$P_2 = P_1(C_2/C_1)^F \times CF$$

$P_2 =$ Required cost of new plant or item.
$C_1 =$ Capacity of plant or item of known cost.
$C_2 =$ Capacity of plant or item of new plant.
$F =$ Exponential factor.
$CF =$ Correction factor for say, inflation.
$P_1 =$ Cost of plant item of known capacity C_1.

"Of course the question is: "Where do we get the factors from?" Answer: historical data."

3.3

During a feasibility study, we may be able to use a factorial cost estimation method[5]. If we have formed a judgement on the market price (our cost) of a main plant item, we can then apply factors for associated costs.

For example: a heat exchanger may cost £20,000, and appropriate factors for the associated costs might be:

Pipework	0.35
Instrumentation	0.15
Civils	0.10
Structural and Building	0.00
Insulation	0.15
Electrical	0.15
Design	0.15
	1.05

Thus the total cost is $= £20,000 \times (1.05 + 1) = £41,000$

---

5 Adapted and taken, with permission, from *A Guide to Capital Cost Estimating*, 3rd edition, published by the Institution of Chemical Engineers and endorsed by the Association of Cost Engineers. Information about the 4th edition can be viewed on the Internet. An excellent little book on the subject.

Again we get the factors from historical data. This also demonstrates that there are 101 small items that add up to as much as the cost of the main item, thus giving credibility to the rule of thumb that when you have 'guestimated' the main components of your estimate, double it.

3.4

Examples of other formulas are for a process plant: count the number of major equipment items and multiply by the number of pipework connections (normally four or five). Then multiply by the average length of a pipework run, and finally multiply by the rate per meter (including valves and instruments).

For standard building work, you need a cost per square meter.

For a journey on the London tube: count the number of stops and multiply by 2.5 minutes (2 minutes on the Piccadilly line). Add five minutes for a change, and you will always complete your project journey on time.

Of course, all of these are qualified by assumptions and exclusions (for example, fees) and then 'it depends on the specification and so on'. But they *all* depend on historical data.

All more-detailed estimates need to be divided into categories, and the four sections listed below show the typical percentages making up an onshore process plant.

		%
a.	Direct Costs:	
	Major equipment	15
	Bulk material	20
	Labour	20
	fabrication	
	construction labour and/or	
	subcontractor installation and/or	
	subcontractor supply and installation	
b.	Indirect Costs:	
	Construction facilities	10
	temporary facilities	
	temporary utilities	
	supporting services	
	construction equipment	
	tools and consumables	
c.	Management Services:	
	engineering, procurement,	
	and construction management	10
d.	Other Elements:	
	Management fee/profit	5
	Escalation	10
	Contingency	10
		___
	Total Project	100

Whilst the above is a model in a particular industry, it is important to understand the proportions in one's own business context.

For example: for standard building work, 50 per cent of the total cost will be the structure and external doors and windows and 50 per cent for internal trades, fixtures, and finishes.

For building renovation work, 50 per cent of the cost will be materials and 50 per cent labour.

4 Estimate Information and Content[6]

The specific allocation of costs elements will depend on the cost coding system used by a particular company.

4.1 The Location/Site

a. Infrastructure purchase (land and buildings and so on), including all associated legal and statutory payments (stamp duty).
b. Site clearance
c. Site survey of ground conditions
d. Investigation into special risks such as earthquakes, likelihood of flooding, and abnormal meteorological conditions
e. Archaeology artefacts
f. Road improvements, diversions, and obstructions, reinforcement of bridges
g. Railway improvements, rail sidings, overhead electrification
h. Road, rail and piping route wayleaves/right of way
i. Dock, jetty, and quayside requirements
j. Water supply – bore holes
k. Sewage and waste disposal
l. Congestion charges, parking fees, fines – for urban projects.

4.2 Main Process Units

a. All process plant and equipment including standby plant
b. Costs of process development and any prototype testing
c. Special erection costs, for example, heavy lifts, special cranes, or clean rooms/conditions
d. Costs due to special materials, refractories
e. Costs due to special manufacturing techniques, glass lined, high pressures, or manufacturing capacity
f. Inspections and tests, such as string tests
g. Delivery costs particularly for heavy, long, oversize, or wide loads and any restrictions or consents required

---

6 Based on and adapted from a Cranfield lecture note by Laurie. F. Williams. – Naturally some of the standard elements of this list appear in *A Guide to Capital Cost Estimating*, 3rd edition, published by the Institution of Chemical Engineers and endorsed by the Association of Cost Engineers.

h. Consumables to be charged as capital, for example, catalysts
i. Safety equipment and any costs associated with a safety incentive scheme
j. Containment of any hazardous operations
k. Fire protection equipment
l. Ventilation for hot conditions, toxic gasses and vapours, dust and fire risks
m. Equipment to meet the requirements of the factory inspector, alkali inspector, and other statutory inspections
n. Effluent treatment plant (including development costs)
o. Instrumentation and controls and development costs
p. Pipework and valves
q. Cathodic protection
r. Insulation and aluminium cladding and painting
s. Mechanical handling facilities
t. Allowance for modifications after erection.

4.3 Off-sites and Utilities

a. Boilers, fired heaters, and steam raising plant and auxiliaries
b. Electricity connection charges
c. Transformers and switchgear
d. Cabling
e. Starters
f. Standby power supplies
g. Plant and pipework for water storage and distribution for process and potable supplies
h. Conventional cooling towers and pipework
i. Raw water clarification plant/water treatment plant
j. Heating, ventilating and air conditioning
k. Lighting conductors
l. Compressed air system
m. Refrigeration, local or centralized
n. Inert or special gas supplies
o. Telephones, communications, and computer costs
p. Test equipment
q. Capital plant spares
r. Cranes, jigs, and maintenance equipment
s. Internal transport for movement and storage of raw materials, intermediate finished products, and associated fuel costs
t. Operating and maintenance manuals, drawings, and so on.

4.4 Civil Works

a. Piling, vibroflotation, and other ground improvements/stabilisation
b. Foundations, special vibration-proof foundations
c. Main entrance road, gate house, and public road modifications
d. Main plant buildings
e. Buildings for service plant

f. Product storage buildings

g. Stores, warehousing, laboratories, workshops, and offices

h. Medical and first-aid centres, fire station

i. Canteen, showers, changing rooms, and lavatories

j. Vehicle maintenance workshops and inspection pits

k. Garages, car parks, and cycle sheds

l. Customs and excise offices and weighbridge

m. Building services

n. Bund walls

o. Storage tank foundations

p. Slip form structures – chimneys, cooling towers

q. Structural steelwork

r. Drainage – surface, chemical, and soil water

s. Pipe and cable ducts

t. Permanent roads and lighting

u. Site security, fencing, clocking stations, and gate houses

v. Land reinstatement, landscaping, and so on.

4.5 Associated Costs

a. Design/Engineering Costs
   i. Process and detailed design
   ii. Use of company associates, local offices
   iii. Purchasing, expediting and inspection
   iv. Use of specialists and consultants
   v. Departmental overheads
   vi. CAD, Design/construction models
   vii. Lloyd's Insurance or special inspection
   viii. Design personnel involved in construction and commissioning
   ix. Travel and living costs.

b. Direct Construction Costs.
   i. Contractor's overheads and profit – OH&P
   ii. Subcontract or direct labour
   iii. Subsidies to labour – transport, changing rooms, canteen meals
   iv. Specialists for rotating equipment – Millwrights
   v. Transport costs – ship charters, air freight
   vi. Overtime working, abnormal weather conditions, local customs, religious holidays and regulations.

c. Temporary Facilities Required for Construction.
   i. Client, project manager, construction manager and site staff offices and furniture
   ii. Temporary power and water supplies
   iii. Ground water pumping
   iv. Communications equipment and computers
   v. Temporary road access and storage areas and associated lighting
   vi. Cost of screening site with hoardings
   vii. Special scaffolding

    viii. Temporary construction workshops (if permanent project workshops unavailable)

    ix. Training School for welding and so on

    x. Site fabrication facilities

    xi. Labour camp and canteen

    xii. Site security, fencing, clocking stations and gate house – (If permanent facilities not available).

d. Overseas Project.

    i. Camp for expatriate construction personnel

    ii. Housing for couples, kitchen equipment and furniture

    iii. Household and food supplies

    iv. Bachelor quarters, furniture and canteen

    v. Social club Swimming pool

    vi. Packing and transport of personal effects.

e. Miscellaneous Overhead Items.

    i. Process or patent fees Agent's fees

    ii. Import/customs duty

    iii. Legal and insurance fees

    iv. Consultants' fees

    v. Tendering costs

    vi. Cost of finance, cost of Bonds

    vii. Proportion of company's research expenditure

    viii. Proportion of company's central administration expenditure

    ix. Cash flow funding/financing cost

    x. Negotiating margin

    xi. Client involvement/interference

    xii. Company's margin/profit.

4.6

In addition to the estimated project cost, other elements are part of the total investment cost, namely the 'owner's costs.'

a. Land acquisition and right of way costs

b. Licenses and royalties fees

c. Feasibility study costs

d. Financing and interest charges during construction

e. Costs for owners supplied project insurances

f. Support duties and taxes

g. Cost of contract and tendering documents

h. Cost for training of operating personnel

i. Owners staff during construction

j. Purchase and transport costs for 'First fill' of the plant

k. Operating spares and mobile equipment – if not part of the main contract

l. Operating supplies and consumables

m. Catalysts and chemicals

n. Commissioning staff costs.

We can thus see that, the more accurate we require the estimate to be, the more data we require and the more work we must do.

5   Contingency Estimation

5.1

In order to arrive at a reasonable contingency, it is useful to divide all the components of the estimate into different categories, depending on the degree of uncertainty.

5.2

As an example, there could be five areas of risk defined as follows (any more are not needed):

a. very low risk
b. low risk
c. medium risk
d. high risk
e. very high risk

5.3

Multiply the proportion of each of the above categories in the total estimate by the agreed contingencies to come up with an overall contingency figure. For example, say the total of items, for which it is considered there is a very low risk, represents 5 per cent of the whole of the initial estimate, the low risk items represent 10 per cent, medium risk items total 20 per cent, high risk items total 45 per cent, and very high-risk items 20 per cent. Assuming that the contingency figures chosen are 2 per cent for very low risk, 5 per cent for low risk, 10 per cent for medium risk, 15 per cent for high risk and 25 per cent for very high risk, then the total contingency calculation would be as follows:

	Breakdown of initial estimate %	Contingency %	Contingency to be added %
Very low risk	5	2	0.10
Low risk	10	5	0.50
Medium risk	20	10	2.00
High risk	45	15	6.75
Very high risk	20	25	5.00
	100		14.35

5.3.1
A total contingency of 14.35 per cent would be added to the estimate. This would be quite a low level of contingency at the start of a project. It would not be unusual to include a total contingency of 20 to 25 per cent or more.

5.4

As the job develops and more items become firm, the percentage of low risk items increases, and the overall contingency should be decreased accordingly. Where the cost of an item has been underestimated in the original budget, funds have to be transferred from contingency to that item. Where the cost of an item has been overestimated, the spare money can be transferred into contingency. A more disciplined approach is to transfer any surplus to the gross margin on the bottom line.

5.5

By regularly reallocating the percentages of the various levels of risk, it is possible to check the original estimate of contingency required.

Section F   Filing and Archiving

This may be a boring subject, but keeping the paper under control is crucial. Ideally a company will have a filing system that is consistent on all projects so that full-time or part-time personnel are familiar with the system regardless of the project. It doesn't matter whether the system is electronic or physical; a system such as this is still required.

1   The Filing System

1.1

The filing system should be designed so as to reduce multiple filing of documents and ensure a complete set of predetermined master files for archiving.

1.2

The system should be based on origin and destination for all correspondence and by subject for all other documents.

1.3

In order to keep the system simple, limit it to three levels, using a decimal numbering system: major categories (1.0.0), minor categories (1.1.0), and specific subjects (1.1.1).

1.4

The major categories should primarily be organized around functional or departmental groups:

1.0.0	Project management
2.0.0	Correspondence/minutes of meetings
3.0.0	Legal and insurance
4.0.0	Finance and accounts
5.0.0	Administration
6.0.0	Project controls/project office
7.0.0	Design/engineering
8.0.0	Procurement
9.0.0	Installation/construction
10.0.0	Commissioning/start-up

1.5

In order to minimise the number of files, it is suggested that subject-specific files are only created as they become necessary. That is, the filing list can be edited down to suit the size of project by omitting unwanted files. However, the standard file numbers should not be changed, e.g. 4.3.3 is always, say, letters of credit – if used.

1.6

Unless there is a central filing system, the master files should be the responsibility of the functional supervisor for the appropriate category. Project management is responsible for the first three major categories. If there is no finance and accounts function, then this becomes the responsibility of project controls or the project office. The administration category is also the responsibility of project controls or the project office.

1.7

The major category master files should contain the original documents or master copies of key documents relating to that subject. They are the official project files and should not be used as working documents.

1.8

When a project group requires copies of material from another major category master file, the copies should be kept in a secondary file, not the group's own major category file.

1.9

The format for secondary files is up to the individuals concerned. However, it would be sensible if they followed the numbering system of the major categories concerned.

1.10

Secondary files are for working convenience and should never contain original documents or master copies. The difficulty is getting people to transfer any marginal comments or relevant working papers to the master file in order to ensure a complete archive record.

1.11

Documents generated within the project should use the subject or minor category as a heading, and the file number should be allocated at the time of generation by the originator.

1.12

Correspondence and action logs should always list the reference number of the relevant subject file.

1.13

In general, documents should only need filing once. The exception is client correspondence and minutes of meetings. These are filed under the appropriate origin–destination file as well as under the relevant subject file category.

1.14

Get company and client approval/agreement and issue a dated list with the revision status.

2   Archiving

2.1

There are three reasons for archiving: (1) statutory requirements; (2) contractual requirements; and (3) as an historical record so that, hopefully, we can learn and transfer know-how for the future.

2.2

Conventionally, archiving occurs at the end of a project. However, consider archiving progressively at key stages of a project if this is possible, particularly whilst key personnel are still available.

2.3

If the master files have been maintained complete, and the secondary files have been kept independent from the master files, then the secondary files can be destroyed at project end. The master files are retained as the project records and can be archived accordingly.

2.4

There are three categories of 'retention', and the file labels should be colour coded accordingly: red, white, and blue (see section 3 following).

2.4.1
Red files should be retained as hard copy up to the end of any project warranty period and can then be transferred to the appropriate archive medium, e.g. DVDs.

2.4.2
Blue files are only retained as hard copy up to the end of the warranty period and can then be destroyed. However, do not forget that the warranty period may, for various reasons, get extended, and you may have to stop admin people from destroying the files prematurely.

2.4.3
White files can be destroyed at project end when the project is handed over to the client.

2.5

The retention of original legal/contract documents should be the responsibility of a company's legal or contracts department. Thus, any original documents of this nature that were not handed over should be transferred before archiving.

2.6

The same principle applies to original insurance documents or other documents, which for legal reasons have to be retained as hard copy. Similarly, all financial documents that are required to be retained for statutory reasons should be transferred to finance and accounts. Thus the project files can be destroyed at the end of the project.

2.7

If material requisition files and purchase order files have been strictly maintained and they are complete with all relevant correspondence and so on, then these files form the prime documents for archiving. Other files, for example internal correspondence files, can be destroyed at the end of the project.

3   Master File Index
Recommended Minor Categories and Suggested Subjects

	1.0.0	**Project Management**
Colour Code	File No.	Title
White	1.1.0	Memos: To and from project management, to and from management, and to and from other key departments
Blue	1.2.0	Coordination procedures: Client coordination procedure, procedures manual, coordination procedure third parties.
White	1.3.0	Policies and personnel: Employment conditions, assignments, recruitment, immigration, work permits, visas, resumes, vacations, and rest and relaxation schedules
White	1.4.0	Public relations: Press releases, presentations, VIP visits
White	1.5.0	Quality assurance: Audits
Red	1.6.0	Commissioning/Handover: Precommissioning, mechanical completion, material certificates, client acceptance certificates, start-up
White	1.7.1	Training: Technology transfer, training client personnel

Colour Code	File No.	Title
Red	1.8.0	Matters of concern.
		When problems occur it is often difficult to track down all the relevant documentation. Consequently, if you, as the project manager, decide that an issue should be highlighted as a 'matter of concern' then a separate secondary subject file is raised under this number. Obviously personnel need to be made aware that a matter of concern has been raised and an extra copy of all the subject documents made available. This should be the only area of multiple filing copies but it saves a lot of hassle if the issue becomes serious.
Hard Copy	1.9.0	Project Historical Report.
	2.0.0	**Correspondence/Minutes of Meetings**

Colour Code	File No.	Title
Red	2.1.0	Client correspondence:
		Letters to and from client home office, to and from site, and to and from another location.
		The second copy of a letter should be copied to the relevant subject file. Letters should be complete with all attachments. Bulky attachments can be filed in a separate letter supplement file, properly cross referenced to the letter file.
Red	2.2.0	Letters to and from client's technical advisor/managing contractor.
Red	2.3.0	Letters to and from company entities.
Red	2.4.0	Letters to and from partners, associates, or local office.
Red	2.5.0	Letters to from finance lenders:
		Letters to and from banks.
Red	2.6.0	Letters to and from consultants/licensors/third parties.
Red	2.7.0	Letters to and from others.
		Letters to or from contractors, subcontractors, or vendors should be filed in the procurement material requisition, purchase order, or contract files as appropriate.
Red	2.8.0	Minutes of meetings with client:
		Filed by subject, e.g. engineering/design, procurement, and installation/construction.
		Alternatively, file by office or location.
Red	2.9.0	Minutes of Meetings with client's technical advisor/managing contractor.
Red	2.10.0	Minutes of meetings with company entities.
Red	2.11.0	Minutes of meetings with partners, associates, or local office.
Red	2.12.0	Minutes of meetings with finance lenders or banks.
Red	2.13.0	Minutes of meetings with consultants/licensors/third parties.
Red	2.14.0	Minutes of meetings with others.
Red	2.15.0	Internal project management minutes of meetings.
white	2.16.0	Internal engineering/design minutes of meetings.
		Meetings with vendors and subcontractors are filed in the appropriate purchase order file.

Colour Code	File No.	Title
	3.0.0	**Legal and Insurance.**
White	3.1.0	Memos: To and from legal/contracts and memos to and from insurance.
White	3.2.0	Tender/proposal: Pre-contract correspondence.
White	3.3.0	Contract (official copy): Contract changes/variations client approved change orders, confidentiality agreements. Remember the originals should always be passed to the corporate legal/contracts department.
Red	3.4.0	Force majeur.
Hard copy	3.5.0	Client job/project specification.
White	3.6.0	Inter-company agreements: Agreements with consultants, partner, and other third parties, associated confidentiality agreements.
Red	3.7.0	Statutory authorities: Planning permission.
White	3.8.0	Insurances: Policies, insurance claims

Colour Code	File No.	Title
	4.0.0	**Finance and Accounts**
White	4.1.0	Memos: Memos to and from finance and accounts.
White	4.2.0	Invoicing: Accounting authorisation and funds/cost flow chart, advance funds requests, fee invoices, cost invoices – home office and site, back charges, interoffice and intercompany.
White	4.3.0	Finance: Banking relationships/lenders, bank accounts, letters of credit, bank guarantees. Correspondence see 2.5.0 All financial records that have to be retained for statutory reasons should be passed to the finance and accounts department
White	4.4.0	Costs: Approval authorities, cash flow, irrecoverable costs, recoverable costs, chart of accounts, foreign exchange, payroll, telephone charges, agency timesheets/invoices, personal accounts, overtime requests. Individual cost files by contractor for materials and services by others. This is usually a site function.
White	4.5.0	Taxes: Taxes and tax laws, manufacturing taxes, income tax, VAT.
White	4.6.0	Payroll: Home office, site, other locations.
Blue	4.7.0	Audit: Client, internal, queries raised.

Colour Code	File No.	Title
Blue		Reports:
		ccounts/receivables due of special significance, payables, multi-loan reports, financial statements, earned revenue calculation, costs and commitments (this will form part of the project manager's report)

5.0.0 **Administration**

Colour Code	File No.	Title
White	5.1.0	Organization charts
White	5.2.0	Project Roster
		Personnel requests & approvals
White	5.3.0	Forms
White	5.4.0	Documentation distribution
		Circulation lists, transmittals.
White	5.4.0	Travel requests
White	5.5.0	Reproduction
White	5.6.0	Leases
White	5.7.0	Furniture and equipment

6.0.0 **Project Controls/Project Office**

Colour Code	File No.	Title
White	6.1.0	Memos:
		Memos to and from project controls/project office.
Blue	6.2.0	Cost estimates:
		Proposal estimate, cost codes, budgets and adjustments, trend base estimate, definitive estimate, updates, cost studies.
		This file should be passed to the main estimating department at project completion for their retention or disposal.
Blue	6.3.0	Cost trends:
		Home office trends, site trends, trends back-up, change orders.
		The master copy of agreed change orders and the associated back-up material must be filed in the appropriate contract sub-file of 3.3.0.
White	6.4.0	Cost forecasts:
		Home office forecasts, site forecasts.
Hard Copy	6.5.0	Reports:
		Monthly progress report, monthly cost report (costs and commitments), non-reimbursable costs report, man hour report, design/engineering progress, critical items, historical cost report.
		The historical cost report forms part of the main project historical report 1.9.0
White	6.6.0	Programmes/Schedules:
		Proposal schedule, project master schedule (original and latest computer update), summary schedules (design/engineering and manufacturing/construction), 30-, 60- or 90-day schedules, detailed schedules (design/engineering and manufacturing/construction), manpower schedule.

Colour Code	File No.	Title
White	6.7.0	Status Reports:
		Material requisitions, major equipment list, outstanding requisitions, status of commitments, critical activities list, traffic/shipping report, other reports.
White	6.8.0	Computer Services:
		Data processing.

7.0.0   Design/Engineering

Colour Code	File No.	Title
White	7.1.0	Memos:
		Memos to and from design/engineering.
White	7.2.0	Administration:
		Personnel and staffing, manpower planning and loading, overtime.
White	7.3.0	Cost and schedule:
		Design/engineering budget and schedule, definitive estimate, overtime.
White	7.4.0	Reports:
		Weekly technical man hours, design group's progress, holds lists, requirements lists.
Red	7.5.0	Process design/engineering.
Blue	7.6.0	Studies.
Red	7.7.0	Design data and calculations
Red	7.8.0	Technical Reports:
		Sub-categories for each design/engineering discipline.
Blue	7.9.0	Specifications.
Red	7.10.0	Data sheets.
Red	7.11.0	Requisitions:
		Material requisitions by number (include all data, correspondence, specifications and so on) material requisition control document.
Red	7.12.0	Drawings:
		Drawing record control file, latest print control record.
Red	7.13.0	Vendor print index:
		Vendor drawings.
Blue	7.14.0	Spares schedules:
		Start-up spares, operational and/or two-year spares.
Blue	7.15.0	Lubrication schedules:
		Flushing oils, operating oils.
Blue	716.0	Special subjects:
		Technical audits.

8.0.0   Procurement

Colour Code	File No.	Title
White	8.1.0	Memos:
		Memos to and from procurement.

Colour Code	File No.	Title
White	8.2.0	Reports:
		Subcontract status report.
White	8.3.0	Vendors:
		Supplier surveys, tender lists, cost and commitment register.
Red		Original copies of tender analyses/summaries.
Red	8.4.0	Purchasing by purchase order:
		Includes the master material requisition.
Red	8.5.0	Inspection by purchase order.
White	8.6.0	Expediting by purchase order.
White	8.7.0	Traffic by purchase order:
		Logistics studies, inspection release certificates, air freight.
Blue	8.8.0	Customs:
		Shipment numbers, bills of lading.
Red	8.9.0	Subcontracts by number:
		Transfer each master file to site after award of the subcontract.
Red	8.10.0	Site requisitions:
		Material requisitions from site requiring home office action.

	9.0.0	**Construction (Home Office)**
Colour Code	File No.	Title
White	9.1.0	Memos:
		Memos to and from home office construction, memos to and from site.
White	9.2.0	Administration:
		Personnel and staffing non-manual, personnel and staffing manual, messing and accommodation.
White	9.3.0	Cost and Schedule:
		Construction budget, definitive estimate, master construction schedule, 30-, 60- or 90-day schedule.
White	9.4.0	Reports:
		Daily/weekly progress, site labour costs, pinch/checklists.
Hard copy		photographs.
Blue	9.5.0	Site Materials:
		Materials requiring home office action.
		Receiving reports, overages – shortages and damage, back-charge claims, warehouse, inventory and disposal of surplus material.
Blue	9.6.0	Subcontracts:
		Subcontract plan, correspondence, minutes of meetings, extras, changes, insurance, warranties, progress payments, claims.
		Sited filed by subcontract number post award.
Blue	9.7.0	Safety:
		Incentive scheme.
Blue	9.8.0	Security.
Blue	9.9.0	Industrial/Labour Relations:
		Site agreement.

Section G   Financial Appraisal[7]

1   Cash versus Profit

It is important to realise the difference between these two concepts. Cash is a liquid asset owned by a business, enabling it to buy goods and services, whereas profit is only an accounting measure of the difference between sales revenue and business expenses. Tabulations in paragraphs 1.1 to 1.3 illustrate a worked example, showing how the two are related. If a project needs *financing*, it usually implies that cash is required.

1.1   Project Parameters

Cost of Project	£150 m
Annual Revenue	£150 m
Annual Costs	£ 50 m
Life of Project	5 Years
End Value	Nil

1.2   Cash Flow

Year:	1	2	3	4	5	6	
Revenue	150	150	150	150	150	0	
Costs	50	50	50	50	50	0	
Capital	150	0	0	0	0	0	
	−50	100	100	100	100	0	Total £350 m
Tax 50%		35	35	35	35	35	
Cash	−50	65	65	65	65	− 35	Total £175 m

1.3   Profit & Loss

Revenue	150	150	150	150	150	0	
Cost	50	50	50	50	50	0	
Depreciation	30	30	30	30	30	0	
	70	70	70	70	70	0	Total £350 m
Tax @ 50%	35	35	35	35	35	0	
	35	35	35	35	35	0	Total £175 m

1.4

It should be noted that not only is the project short of (£50 m) cash to finance the project but also, (since the accountants have made a profit) needs £35 m cash to pay the taxman and a further £35 m has to be found to pay the shareholders a dividend.

---

7 This Section is taken with permission from Professor David Middleton's book *Financial Decisions*.

1.5

Other costs can also be manipulated. On one occasion the head of the management school said to me: "Your MSc course never makes any profit." Fortunately, I had my wits about me and responded with the fact that the school overheads would be virtually unchanged without my course and responded with: "It depends on how much overhead you want to load onto my course." This scenario is valid for most business environments and is particularly true of manufacturing. In a project environment, the overhead will usually be fixed at an hourly rate right across the board. However, the amount held in reserve for various risks on a project can be adjusted as you wish.

1.5.1

One American company I worked for never made a profit. The senior vice president would visit once a year, spend the morning in the accounts department, and then have lunch with the UK general manager. During the meal, advice would be given about the strategy to be adopted for the following year, and at the end of the meal an invoice for consultancy services would be presented. The invoice amount was, coincidentally, the same as the profit that had been determined in the morning. Why? The company did not want to pay taxes in the UK but preferred to pay taxes in the U.S.A.

1.6

Lenders of money will consider several factors, listed below, before cash is advanced:

a. Security   If the project fails, will they still get their money back?
b. Inflation   Will the return of the money cover for future inflation?
c. Timing   When will the money be repaid?
d. Return   Will there be a return over and above inflation to cover for the risk?
e. Opportunity Cost   Would it be better to lend the money to another project?

1.7

Project appraisal usually involves looking at the returns of a project in terms of cash. Even if the project is being undertaken for other than financial reasons, the benefits are often given cash values and the scheme viewed exclusively from an accounting viewpoint. This system is called *cost–benefit analysis*. Projects usually involve the spending of cash now with a view to getting more cash back at a later date – see Figure V.G.1 which shows a typical cash curve for a project.

A-B	Conceptual	E-F	Start-up
B-C	Planning/design	F-G	Operation
C-D	Implementation	G-H	Run down
D-E	Commissioning	H-I	Decommission

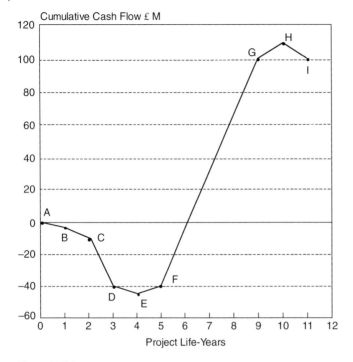

Figure V.G.1

2 Simple Project Appraisal Methods

2.1 Average Rate of Return on Investment

Here the cash flows are added up over the project life and an annual return calculated. This figure divided by the initial investment gives the average rate of return; see the tabulations below.

		Project A (£'000)	Project B (£'000)
Investment	Year 0	−6	−12
Cash Inflows	Year 1	+3	+7
	Year 2	+4	+8
	Year 3	+8	+9
1. Total cash inflows		+15	+24
2. Total net profit		+9	+12
3. Average annual profit		+3	+4
4. $\dfrac{\text{Average annual profit}}{\text{Initial investment}}$		$\dfrac{3}{6} = 50\%$	$\dfrac{4}{12} = 33\%$

The average rate of return on investment (expressed as a percentage) tells us something about the profitability of a capital project. However, the averaging process eliminates relevant information about the timing of the cash flows.

		Project C (£'000)	Project D (£'000)
Investment	Year 0	−6	−6
	Year 1	+1	+6
Cash Inflows	Year 2	+2	+2
	Year 3	+6	+1
1. Total net profits		+9	+9
2. Average annual profit		+3	+3

In the second example Project D is preferable since the additional cash received in year 1 can be invested – there is an additional *opportunity cost*. However, the simple method of calculating the average rate of return on investment does not indicate this, and it ignores the timing of the returns.

3   Payback

This method simply identifies the time for the returns from a project to exceed the initial outlay. Projects with a shorter payback are given preference as the cash exposure (and hence the risk) are minimised. The key problem with this method is that it completely ignores all cash flows after payback see Figure V.G.2 below.

		Project E (£'000)	Project F (£'000)
Investment	Year 0	−6	−6
	Year 1	+3	+8
Cash Inflows	Year 2	+4	+4
	Year 3	+8	+3

3.1

Payback Periods for Project E and Project F (see next page.)

4   Discounted Cash Flow Techniques

These methods take into account not only the amount but also the timing of all cash flows over a project's lifecycle. They do this by identifying the difference in the positive and negative cash flows for a project for each year of its life and then *discounting* these amounts back to the present. If the net present value (NPV) of these cash flows is positive, then the project is viable, and if negative, it is not.

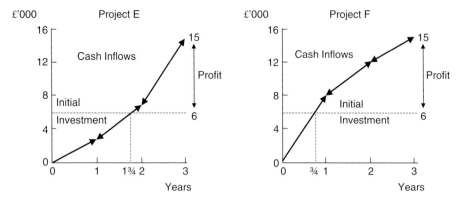

Figure V.G.2

4.1

Let us suppose that money can be invested today to yield 10 per cent a year; then £100 invested today will accumulate (compounding annually) to the amounts shown below:

In 1 year's time, to £1 00 × 1. 1 0 = £100 × 1.100 = £110.00
In 2 years' time, to £100 × (1.10)² = £100 × 1.210 = £121.00
In 3 years' time, to £100 × (1.10)³ = £100 × 1.331 = £133.10.

	End of Year 0* (£)	End of Year 1 (£)	End of Year 2 (£)	End of Year 3 (£)
Future values	100.00 ——	110.00 ——	121.00 ——	133.10
Present Values {	100.00 ——	110.00		
	100.00 ————————————		121.00	
	100.00 ————————————————————			133.10

* The end of Year 0 is 'the present' or simply, now.

4.2

What, then, is the present value of £100.00 to be received at the end of three years? Clearly it must be £75.13, as shown below:

$$\frac{£100.00}{(1.10)^3} = \frac{£100.00}{1.331} \left[or \frac{£100.00}{£133.10} \times £100.00 \right] = £75.13$$

We can prove this, by showing what would happen if we invested £75.13 at 10 per cent a year. Each year the effect of compound interest is to add 10 per cent of the start-of-year amount to the cumulative amount invested:

After 1 year the amount becomes: £75.13 + £7.51 = £ 82.64
After 2 years the amount becomes: £82.64 + £8.27 = £ 90.91
After 3 years the amount becomes: £90.91 + £9.09 = £100.00

Accordingly, the present value of £100 to be received at a future date is shown below:

	End of Year 0* (£)	End of Year 1 (£)	End of Year 2 (£)	End of Year 3 (£)
Future values	75.13	82.64	90.91	100.00
Present Values	90.91 ——— 82.64 ———————— 75.13 ————————————	100.00	100.00	100.00

4.3   Net Present Value – NPV

The net present value calculates a project's return by comparing cash payments and cash receipts at the same point in time by discounting future cash flows back to the present

End of year (EOY)	Cash flows (£)	Discount Factor (at 10%)	'Present' EOY 0 value (£)	(£)
0	$-10{,}000\ [\div(1.10)^0] \times$	$1.000 =$	$-10{,}000$	$= -10{,}000$
1	$+3{,}000\ [\div(1.10)^1] \times$	$0.909 =$	$+2{,}727$	
2	$+4{,}000\ [\div(1.10)^2] \times$	$0.827 =$	$+3{,}308$	$= +9{,}790$
3	$+5{,}000\ [\div(1.10)^3] \times$	$0.751 =$	$+3{,}755$	
		Net Present Value		$= -210$

Even ignoring inflation, £10 now is worth more than £10 in a year's time because of opportunity costs. Thus it can be seen that money has a *time value,* and this can be given a value by applying a discounting factor to the cash flows. The tabulations, in paragraphs 4.2 and 4.3 above, demonstrate this technique. The calculation of these discounted amounts can be done by use of standard discount tables (see Table A and Table B at the end of this section). Nowadays, however, most spreadsheet programs enable even very complex NPV calculations to be undertaken on personal computers.

4.4

If the discount factor used includes a figure for future inflation, then it is important to forecast the cash flows allowing for inflation. If the discount factor ignores inflation, then so, too, must the cash flows. The discount factor (sometimes called hurdle or criterion rate) is usually unique to each company or organization and is often updated each year. Its value depends upon a highly complex calculation, which establishes the company's cost of borrowing from banks, shareholders, and so on.

5 Internal Rate of Return – IRR

The internal rate of return is simply that discount rate, which if applied to the project cash flows, results in an NPV of approximately zero. The tabulation below shows a project where the IRR is 8.9 per cent (see Figure V.G.3). The higher the IRR is over the company discount rate, the more financially robust the project is seen to be.

EOY	Cash flows (£)		Discount Factor (at 9%)	Present value (£)	(£)
0	−10,000		1.000 =	−10,000	= −10,000
1	+3,000 [÷ $(1.09)^1$] ×		0.917 =	+2,751	
2	+4,000 [÷ $(1.09)^2$] ×		0.842 =	+3,368	= + 9,979
3	+5,000 [÷ $(1.09)^3$] ×		0.772 =	+3,860	
			Net present value		= −21

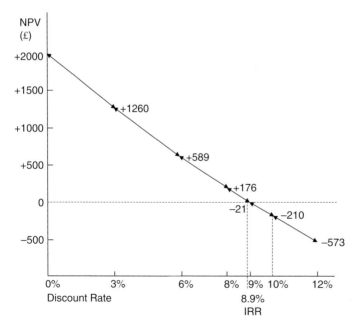

Figure V.G.3

6 Sensitivity and Risk Analysis

Once the project financial model is loaded on a PC system, it is very easy to change underlying assumptions one at a time and see what effect this will have on the NPV and IRR. The resulting plot of 'straight' lines in the form of a star burst emanating from the origin shows the sensitivity of the project to these changing variables.

6.1

What is interesting is that changing the cost of design/engineering has minimal impact on the NPV of the project. Whereas, changing sales data for example, unit sales price, by say 5 or 10 per cent, has a very significant effect on the viability of the project. The salesperson shrugs their shoulders and says, 'It is difficult to forecast the marketplace' whilst the design manager could get the sack!

6.2

Whilst this plot is very useful, it does not show the likelihood of these assumptions actually changing. To overcome this problem, it is possible to give probability values to each assumption made and then summate all these probabilities to give an overall project plot.

7 Financial Appraisal Conclusion

It is important when assessing a project to use as many different appraisal techniques as possible. Each method has its weaknesses but, used together, they can provide a very good overall financial picture. Having said this, it is vital not to get too carried away with pure numerical analysis, most projects involve non-financial factors and in some cases these may be more important than the 'bottom line.'

Table A Present Value of £1

Years Hence	1%	2%	4%	6%	8%	10%	12%	14%	15%	16%	18%	20%	22%	24%	25%	26%	28%	30%	35%	40%	45%	50%
1.......	0.990	0.980	0.962	0.943	0.926	0.909	0.893	0.877	0.870	0.862	0.847	0.833	0.820	0.806	0.800	0.794	0.781	0.769	0.741	0.714	0.690	0.667
2.......	0.980	0.961	0.925	0.890	0.857	0.826	0.797	0.769	0.756	0.743	0.718	0.694	0.672	0.650	0.640	0.630	0.610	0.592	0.549	0.510	0.476	0.444
3.......	0.971	0.942	0.889	0.840	0.794	0.751	0.712	0.675	0.658	0.641	0.609	0.579	0.551	0.524	0.512	0.500	0.477	0.455	0.406	0.364	0.328	0.296
4.......	0.961	0.924	0.855	0.792	0.735	0.683	0.636	0.592	0.572	0.552	0.516	0.482	0.451	0.423	0.410	0.397	0.373	0.350	0.301	0.260	0.226	0.198
5.......	0.951	0.906	0.822	0.747	0.681	0.621	0.567	0.519	0.497	0.476	0.437	0.402	0.370	0.341	0.328	0.315	0.291	0.269	0.223	0.186	0.156	0.132
6.......	0.942	0.888	0.790	0.705	0.630	0.564	0.507	0.456	0.432	0.410	0.370	0.335	0.303	0.275	0.262	0.250	0.227	0.207	0.165	0.133	0.108	0.088
7.......	0.933	0.871	0.760	0.665	0.583	0.513	0.452	0.400	0.376	0.354	0.314	0.279	0.249	0.222	0.210	0.198	0.178	0.159	0.122	0.095	0.074	0.059
8.......	0.923	0.853	0.731	0.627	0.540	0.467	0.404	0.351	0.327	0.305	0.266	0.233	0.204	0.179	0.168	0.157	0.139	0.123	0.091	0.068	0.051	0.039
9.......	0.914	0.837	0.703	0.592	0.500	0.424	0.361	0.308	0.284	0.263	0.225	0.194	0.167	0.144	0.134	0.125	0.108	0.094	0.067	0.048	0.035	0.026
10.......	0.905	0.820	0.676	0.558	0.463	0.386	0.322	0.270	0.247	0.227	0.191	0.162	0.137	0.116	0.107	0.099	0.085	0.073	0.050	0.035	0.024	0.017
11.......	0.896	0.804	0.650	0.527	0.429	0.350	0.287	0.237	0.215	0.195	0.162	0.135	0.112	0.094	0.086	0.079	0.066	0.056	0.037	0.025	0.017	0.012
12.......	0.887	0.788	0.625	0.497	0.397	0.319	0.257	0.208	0.187	0.168	0.137	0.112	0.092	0.076	0.069	0.062	0.052	0.043	0.027	0.018	0.012	0.008
13.......	0.879	0.773	0.601	0.469	0.368	0.290	0.229	0.182	0.163	0.145	0.116	0.093	0.075	0.061	0.055	0.050	0.040	0.033	0.020	0.013	0.008	0.005
14.......	0.870	0.758	0.577	0.442	−0.340	0.263	0.205	0.160	0.141	0.125	0.099	0.078	0.062	0.049	0.044	0.039	0.032	0.025	0.015	0.009	0.006	0.003
15.......	0.861	0.743	0.555	0.417	0.315	0.239	0.183	0.140	0.123	0.108	0.084	0.065	0.051	0.040	0.035	0.031	0.025	0.020	0.011	0.006	0.004	0.002
16.......	0.853	0.728	0.534	0.394	0.292	0.218	0.163	0.123	0.107	0.098	0.071	0.054	0.042	0.032	0.028	0.025	0.019	0.015	0.008	0.005	0.003	0.002
17.......	0.844	0.714	0.513	0.371	0.270	0.198	0.146	0.108	0.093	0.080	0.060	0.045	0.034	0.026	0.023	0.020	0.015	0.012	0.006	0.003	0.002	0.001
18.......	0.836	0.700	0.494	0.350	0.250	0.180	0.130	0.095	0.081	0.069	0.051	0.038	0.028	0.021	0.018	0.016	0.012	0.009	0.005	0.002	0.001	0.001
19.......	0.828	0.686	0.475	0.331	0.232	0.164	0.116	0.083	0.070	0.060	0.043	0.031	0.023	0.017	0.014	0.012	0.009	0.007	0.003	0.002	0.001	
20.......	0.820	0.673	0.456	0.312	0.215	0.149	0.104	0.073	0.061	0.051	0.037	0.026	0.019	0.014	0.012	0.010	0.007	0.005	0.002	0.001	0.001	
21.......	0.811	0.660	0.439	0.294	0.199	0.135	0.093	0.064	0.053	0.044	0.031	0.022	0.015	0.011	0.009	0.008	0.006	0.004	0.002	0.001		
22.......	0.803	0.647	0.422	0.278	0.184	0.123	0.083	0.056	0.046	0.038	0.026	0.018	0.013	0.009	0.007	0.006	0.004	0.003	0.001	0.001		
23.......	0.795	0.634	0.406	0.262	0.170	0.112	0.074	0.049	0.040	0.033	0.022	0.015	0.010	0.007	0.006	0.005	0.003	0.002	0.001			
24.......	0.788	0.622	0.390	0.247	0.158	0.102	0.066	0.043	0.035	0.028	0.019	0.013	0.008	0.006	0.005	0.004	0.003	0.002	0.001			
25.......	0.780	0.610	0.375	0.233	0.146	0.092	0.059	0.038	0.030	0.024	0.016	0.010	0.007	0.005	0.004	0.003	0.002	0.001	0.001			
26.......	0.772	0.598	0.361	0.220	0.135	0.084	0.053	0.033	0.026	0.021	0.014	0.009	0.006	0.004	0.003	0.002	0.002	0.001				
27.......	0.764	0.586	0.347	0.207	0.125	0.076	0.047	0.029	0.023	0.018	0.011	0.007	0.005	0.003	0.002	0.002	0.001	0.001				
28.......	0.757	0.574	0.333	0.196	0.116	0.069	0.042	0.026	0.020	0.016	0.010	0.006	0.004	0.002	0.002	0.002	0.001	0.001				
29.......	0.749	0.563	0.321	0.185	0.107	0.063	0.037	0.022	0.017	0.014	0.008	0.005	0.003	0.002	0.002	0.001	0.001	0.001				
30.......	0.742	0.552	0.308	0.174	0.099	0.057	0.033	0.020	0.015	0.012	0.007	0.004	0.003	0.002	0.002	0.001	0.001					
40.......	0.672	0.453	0.208	0.097	0.046	0.022	0.011	0.005	0.004	0.003	0.001	0.001										
50.......	0.608	0.372	0.141	0.054	0.021	0.009	0.003	0.001	0.001	0.001												

Table B Present Value of £1 received annually for N years

Years (N)	1%	2%	4%	6%	8%	10%	12%	14%	15%	16%	18%	20%	22%	24%	25%	26%	28%	30%	35%	40%	45%	50%
1.........	0.990	0980	0.962	0.943	0.926	0.909	0.893	0.877	0.870	0.862	0.847	0.833	0.820	0.806	0.800	0.794	0.781	0.769	0.741	0.714	0.690	0.667
2.........	1.970	1.942	1.886	1.833	1.783	1.736	1.690	1.647	1.626	1.605	1.566	1.528	1.492	1.457	1.440	1.424	1.392	1.361	1.289	1.224	1.165	1.111
3.........	2.941	2.884	2.775	2.673	2.577	2.487	2.402	2.322	2.283	2.246	2.174	2.106	2.042	1.981	1.952	1.923	1.868	1.816	1.696	1.589	1.493	1.407
4.........	3.902	3.808	3.630	3.465	3.312	3.170	3.037	2.914	2.855	2.798	2.690	2.589	2.494	2.404	2.362	2.320	2.241	2.166	1.997	1.849	1.720	1.605
5.........	4.853	4.713	4.452	4.212	3.993	3.791	3.605	3.433	3.352	3.274	3.127	2.991	2.864	2.745	2.689	2.635	2.532	2.436	2.220	2.035	1.876	1.737
6.........	5.795	5.601	5.242	4.917	4.623	4.355	4.111	3.889	3.784	3.685	3.498	3.326	3.167	3.020	2.951	2.885	2.759	2.643	2.385	2.168	1.983	1.824
7.........	6.728	6.472	6.002	5.582	5.206	4.868	4.564	4.288	4.160	4.039	3.812	3.605	3.416	3.242	3.161	3.083	2.937	2.802	2.508	2.263	2.057	1.883
8.........	7.652	7.325	6.733	6.210	5.747	5.335	4.968	4.639	4.487	4.344	4.078	3.837	3.619	3.421	3.329	3.241	3.076	2.925	2.598	2.331	2.108	1.922
9.........	8.566	8.162	7.435	6.802	6.247	5.759	5.328	4.946	4.772	4.607	4.303	4.031	3.786	3.566	3.463	3.366	3.184	3.019	2.665	2.379	2.144	1.948
10.........	9.471	8.983	8.111	7.360	6.710	6.145	5.650	5.216	5.019	4.833	4.494	4.192	3.923	3.682	3.571	3.465	3.269	3.092	2.715	2.414	2.168	1.965
11.........	10.368	9.787	8.760	7.887	7.139	6.495	5.937	5.453	5.234	5.029	4.656	4.327	4.035	3.776	3.656	3.544	3.335	3.147	2.752	2.438	2.185	1.977
12.........	11.255	10.575	9.385	8.384	7.536	6.814	6.194	5.660	5.421	5.197	4.793	4.439	4.127	3.851	3.725	3.606	3.387	3.190	2.779	2.456	2.196	1.985
13.........	12.134	11.343	9.986	8.853	7.904	7.103	6.424	5.842	5.583	5.342	4.910	4.533	4.203	3.912	3.780	3.656	3.427	3.223	2.799	2.468	2.204	1.990
14.........	13.004	12.106	10.563	9.295	8.244	7.367	6.628	6.002	5.724	5.468	5.008	4.611	4.265	3.962	3.824	3.695	3.459	3.249	2.814	2.477	2.210	1.993
15.........	13.865	12.849	11.118	9.712	8.559	7.606	6.811	6.142	5.847	5.575	5.092	4.675	4.315	4.001	3.859	3.726	3.483	3.268	2.825	2.484	2.214	1.995
16.........	14.718	13.578	11.652	10.106	8.851	7.824	6.974	6.265	5.954	5.669	5.162	4.730	4.357	4.033	3.887	3.751	3.503	3.283	2.834	2.489	2.216	1.997
17.........	15.562	14.292	12.166	10.477	9.122	8.022	7.120	6.373	6.047	5.749	5.222	4.775	4.391	4.059	3.910	3.771	3.518	3.295	2.840	2.492	2.218	1.998
18.........	16.398	14.992	12.659	10.828	9.372	8.201	7.250	6.467	6.128	5.818	5.273	4.812	4.419	4.080	3.928	3.786	3.529	3.304	2.844	2.494	2.219	1.999
19.........	17.226	15.678	13.134	11.158	9.604	8.365	7.366	6.550	6.198	5.877	5.316	4.844	4.442	4.097	3.942	3.799	3.539	3.311	2.848	2.496	2.220	1.999
20.........	18.046	16.351	13.590	11.470	9.318	8.514	7.469	6.623	6.259	5.929	5.353	4.870	4.460	4.110	3.954	3.808	3.546	3.316	2.850	2.497	2.221	1.999
21.........	18.857	17.011	14.029	11.764	10.017	8.649	7.562	6.687	6.312	5.973	5.384	4.891	4.476	4.121	3.963	3.816	3.551	3.320	2.852	2.498	2.221	2.000
22.........	19.660	17.658	14.451	12.042	10.201	8.772	7.645	6.743	6.359	6.011	5.410	4.909	4.488	4.130	3.970	3.822	3.556	3.323	2.853	2.498	2.222	2.000
23.........	20.456	18.292	14.857	12.303	10.371	8.883	7.718	6.792	6.399	6.044	5.432	4.925	4.499	4.137	3.976	3.827	3.559	3.325	2.854	2.499	2.222	2.000
24.........	21.243	18.914	15.247	12.550	1C.529	8.985	7.784	6.835	6.434	6.073	5.451	4.937	4.507	4.143	3.981	3.831	3.562	3.327	2.855	2.499	2.222	2.000
25.........	22.023	19.523	15.622	12.783	10.675	9.077	7.843	6.873	6.464	6.097	5.467	4.948	4.514	4.147	3.985	3.834	3.564	3.329	2.356	2.499	2.222	2.000
26.........	22.795	20.121	15.983	13.003	10.810	9.161	7.896	6.906	6.491	6.118	5.480	4.956	4.520	4.151	3.988	3.837	3.566	3.330	2.856	2.500	2.222	2.000
27.........	23.560	20.707	16.330	13.211	10.935	9.237	7.943	6.935	6.514	6.136	5.492	4.964	4.524	4.154	3.990	3.839	3.567	3.331	2.356	2.500	2.222	2.000
28.........	24.316	21.281	16.663	13.406	11.051	9.307	7.984	6.961	6.534	6.152	5.502	4.970	4.528	4.157	3.992	3.840	3.568	3.331	2.857	2.500	2.222	2.000
29.........	25.066	21.844	16.984	13.591	11.158	9.370	8.022	6.983	6.551	6.166	5.510	4.975	4.531	4.159	3.994	3.841	3.569	3.332	2.857	2.500	2.222	2.000
30.........	25.808	22.396	17.292	13.765	11.258	9.427	8.055	7.003	6.566	6.177	5.517	4.979	4.534	4.160	3.995	3.842	3.569	3.332	2.857	2.500	2.222	2.000
40.........	32.835	27.355	19.793	15.046	˥.925	9.779	8.244	7.105	6.642	6.234	5.548	4.997	4.544	4.166	3.999	3.846	3.571	3.333	2.857	2.500	2.222	2.000
50.........	39.196	31.424	21.482	15.762	12.234	9.915	8.304	7.133	6.661	6.246	5.554	4.999	4.545	4.167	4.000	3.846	3.571	3.333	2.857	2.500	2.222	2.000

Section H   Incoterms®

Incoterms are: the ICC (International Chamber of Commerce) official rules for the interpretation of trade terms applying to the contract of sale. The latest terms published in September 2010 came into force 1 January 2011. In the introduction to the earlier 2000 edition of this excellent booklet, it states: 'The purpose of Incoterms is to provide a set of international rules for the interpretation of the most commonly used trade terms in foreign trade. Thus, the uncertainties of different interpretations of such terms in different countries can be avoided or at least reduced to a considerable degree.'

Incoterms tend to be revised and reissued every ten years, reflecting the problems that buyers and sellers have experienced over the preceding period. Older terms can be used provided the purchase contract clearly states the Incoterms year of issue.

The eleven[8] 2010 rules are presented in two distinct groups. Group 1 is for any mode(s) of transport, including maritime transport. Group 2 is for sea and inland waterway transport only.

1   Rules for Any Mode or Modes of Transport

EXW	Ex Works	
FCA	Free Carrier	Goods cleared for export
CPT	Carriage Paid to	FCA + Contract for carriage paid
CIP	Carriage and Insurance Paid to	CPT + Cost of insurance
DAP	Delivered at Place	All costs and risks paid for
DAT	Delivered at Terminal	DAP + Cost of unloading at normal point
DDP	Delivered Duty Paid	DAP + Paying clearance and duty

1.1

This first group of seven rules can be used irrespective of the mode of transport selected and irrespective of whether one or more than one mode of transport is used.

1.2

These rules can be used even when there is no maritime transport at all. Nevertheless, it is important to remember that these rules *can* be used in cases where a ship *is* used for part of the carriage.

---

8 The number of Incoterms rules have been reduced from 13 (Incoterms 2000) to 11. Four terms have been deleted from the Incoterms 2000 rules (DAF, DES, DEQ & DDU), and two new terms have been added, irrespective of the mode of transport, namely: DAT – Delivered at Terminal and DAP – Delivered at Place.

2 Rules for Sea and Inland Waterway Transport

FAS	Free Alongside Ship	
FOB	Free on Board	
CFR	Cost and Freight	FOB + Contract of carriage to named destination
CIF	Cost Insurance and Freight	CFR + Payment of insurance

2.1

In the second group of four rules, the point of delivery and the place to which the goods are carried to, for the buyer, are *both* ports. Hence the label *sea and inland waterway* rules.

2.2

Under the last three Incoterms rules, all mention of the ship's rail as the point of delivery (Incoterms 2000 rules) has been omitted in preference for the goods being delivered when they are *on board* the vessel. This more closely reflects modern commercial reality. It avoids what is considered as the rather dated image of the risk swinging to and fro across an imaginary perpendicular line above the ship's rail.

3 Transfer of Risks and Obligations

3.1

EXW represents the seller's minimum obligation to make the goods available at the seller's premises.

3.2

With the 'F' terms, the seller has extended obligations to hand over goods for carriage free of risk and expense to a carrier nominated by the buyer (FCA, FAS, and FOB).

3.3

With the 'C' terms, the seller has extended obligations to hand over goods for carriage free of risk and expense to a carrier chosen and paid for by the seller (CFR and CPT), together with insurance risks in transit (CIF and CIP).

   With the latter two terms, it should be noted that risk passes when the goods are on a vessel, even though the seller pays the cost and insurance to the named port of destination. That is, the seller must bear costs even after the critical point for the division of risk has happened.

3.4

DAT, DAP, and DDP represent the seller's maximum obligation to deliver at the destination.

4   Sellers' and Buyers' Detailed Obligations

Each term is defined, and the seller's obligations are listed under a series of consistent paragraphs A1 to A10

4.1

A1	General obligations
A2	Licences, authorizations, and formalities
A3	Contracts of carriage and insurance
A4	Delivery
A5	Transfer of risks
A6	Allocation of costs
A7	Notice to the buyer
A8	Delivery documents
A9	Checking, packaging, inspection, and marking
A10	Assistance with information and related costs

4.2

The above terms are matched on a facing page, listing the buyer's obligations in similar paragraphs B1 to B10.

B1	Payment of the price
B2	as above
B3	as above
B4	Taking delivery
B5	as above
B6	as above
B7	Notice to the seller
B8	as above
B9	Inspection of the goods
B10	as above

5   Additional Information

5.1

Be wary of U.S. suppliers, since they signed up to Incoterms relatively late in life. Some suppliers may still believe that FOB means on a truck in the middle of Texas! Consequently, always be pedantic and quote 'Incoterms 2010,' depending on the version you want to use or whatever is the latest issue date.

5.2

If you need more information borrow the Incoterms book from your purchasing department.

Section J Joint Associations

A joint association is an enterprise formed by two or more undertakings for the purpose of executing a particular, clearly defined objective. The joint association will have a single contract with the owner and will be *jointly and severally liable* for the risks involved. Typical arrangements are:

- Ad hoc partnering
- Prime and subcontractor
- Single project contractual joint venture
- Multi-project contractual joint venture
- Partnerships
- Joint venture company (joint stock company)
- Broadly based strategic alliance

In a consortium, each company has a separate contract with the owner. It is, therefore, important to build the safety case into each of the consortium agreements. Consequently, it is not strictly a joint association.

An alliance, however, is a short-term strategy applied to a particular project for the purposes of sharing risk and reward, whereas partnering is a long-term, mutually beneficial contractual arrangement between two or more parties, with compensation linked to performance objectives. The whole essence of partnering arrangements is a realignment of interests. These collaborative relationships involve an approach, which is focused on teamwork and incentives.

International associations are more complex, more costly, and time consuming. They have different cultures and systems and require sophisticated risk management.

International associations are often used on the basis that the local partner will help to bring in work. Therefore, consideration should be given to forming a separate company for tax purposes.

1 Reasons for Joint Association

1.1

Joint associations are used in order to produce *synergy* by combining the skills of the two companies involved. They are a means by which firms can enhance their competitive positions. In effect achieving more together, than the two companies on their own.

1.2

There are not many, if any, advantages for the client. The contractors' reasons are:

a. Mega project
b. Risk sharing
c. Combination of resources
d. Technology transfer
e. Local politics or legislation (Making facilitation payments)
f. Reduce the tender list
g. Taxation
h. Research and development

2 Documentation and Legal Requirements

2.1

An initial protocol or memorandum of understanding will be needed long before the enquiry is received.

2.2

You must have a signed document. However, you won't want to declare it in case the client puts someone else on the tender list instead.

2.3

It is better to have separate agreements for different objectives rather than one big contract agreement. Agreements needed are:

a. Joint venture agreement
b. Joint stock company. Some countries insist on being a company for tax purposes. If you are a company, do you incorporate locally or offshore? A shareholder's agreement will be needed.
c. Articles of association
d. Bylaws
e. Seconding agreements
f. Services agreement
g. Technology transfer of licensing agreements
h. Business plan
i. Licenses and registrations
j. Agreement for doing work back home
k. Association directives; see 6.12 below.

3 Selecting a Partner

3.1

The compatible goals can be summarised as 'half a cake is better than none.' However, shared and attainable goals are essential criteria for success. Goals may only be congruent up until the moment the contract is signed. From that point onwards the partners' objectives will diverge.

3.2

One partner will usually have a technical capability that the other partner does not have – and vice versa. They then spend their time trying to acquire that capability from each other! Be wary of a prospective partner negotiating in order to obtain data without any intention of fulfilling a deal.

3.3

One partner will invariably be dominant in terms of financial strength.

3.4

Ideally the two parties should have common corporate values, with compatible cultures. Without top management commitment, the venture is doomed. Compatibility and commitment are essential elements for success.

3.5

Set objectives for the alliance. Collaborate with your competitors – and win.

3.5.1

However, before forming a joint venturem look for a simpler way. Ask: "Why can't the work be subcontracted?" See Part IV, Section R Subcontracting.

4 Joint Association Risks

4.1

One has to ask, 'Are they a good way of sharing and reducing risks or a way of creating new risks?' It is not half the risk because there are two companies. 'Jointly and severally liable' means taking on the liability for the partner's risks. If there is a problem, the client will go for the partner with the strongest balance sheet. Further, if the joint venture company thinks it is your fault, you may find it difficult getting money from your partners. The risks are:

a. Additional risks to normal contract risks.
b. Turnkey liability without total control.
c. Increased risk size compared to the company size.
d. The company's risk capacity is reduced.

4.2

You are in fact training a competitor!

5 Steps to Evaluate Joint Associations

Addressing cultural issues helps ensure that the changes are permanent, and the tendency to revert to 'business as usual' is minimized.

5.1

Establish the objectives and priorities of your own company.

5.2

Gain and maintain commitment and support from top management.

5.3

Prepare budgets and commit resources.

5.4

Allocate sufficient time.

5.5

Establish a negotiating strategy.

5.6

Follow it through.

5.7

Resolve all the key issues (see below) before entering joint associations. They must all be established *before* the prospect of the project is evident.

5.8

Do not take shortcuts. There is no substitute for clear agreements and documentation. Marry in haste – repent at leisure!

5.9

Do not get the cart before the horse.

5.10

Complete the documentation and legal formalities.

5.11

Commence operation in an orderly fashion. It will take time and effort to get systems and procedures (for say, cost control, man hours, and so on) working.

5.12

Re-evaluate the joint operation when the project is nearing completion.

6 Key Issues for a Joint Association

6.1

The agreement which will set out the association's undertaking to work together for common objectives will normally include:

a. The target cost estimate (the contractual budget)
b. The acceptance criteria for schedule and quality
c. The sharing of influence over the risk versus reward criteria
d. The sharing of the benefits of the risk and reward scheme

6.2

Define the technical and geographical scope for each entity.

6.3

Agree which party will have overall control and define the contribution of each party. Comparable contributions are another essential for success.

6.4

Agree the pricing components and what is or is not included in overheads. How do you know that the other party has given you their best price?

6.5

Agree the capital contributions and financial support provided by each party and the procedure for access to financing.

6.6

Define how financial control will be exercised. See 9.0 below.

6.7

Define how licensing or technology transfer will operate.

6.8

Make sure that the other party is committed exclusively to your company. Only work with each other on the type of work to be performed.

6.9

Agree how you will terminate and disengage from the association. In other words, how will you get divorced? How will you carve up the cake when you separate?

6.10

Create a mechanism for dispute resolution. A 'gin and tonic' clause

6.11

Create a management organization structure involving senior managers from each party – a project board or steering committee. Its purpose is to:

a. Reinforce corporate commitment to the association
b. Promote association principles
c. Review cost and schedule progress
d. Resolve disputes which cannot be settled at project level
e. Aid the resolution of external issues
f. Encourage maintenance of association behaviour

6.11.1

The project board or steering committee should meet at least once a month to see how the joint association is working. Don't let them worry about detail, but concentrate on issues that might be causing problems. The meeting could take half a day. It takes a lot of effort. Don't underestimate the time and effort required.

6.12

A set of association directives will be needed. You need internal procedures that are well established for:

a. The type of work to be performed.
b. What other projects can be tendered.
c. How risks will be shared.
d. Establishing signature authorities and delegation limits.

7   Steps in Tendering

a. Identify a prospect.
b. Identify a partner (before the enquiry does).
c. Develop heads of agreement.
d. Prequalify.
e. Receive an enquiry.
f. Sign a pretender agreement.
g. Sign a preliminary joint association agreement.
h. Submit the proposal or tender.
i. Revise the joint association agreement.
j. Sign the contract.

8 Control of the Work

8.1

For the joint association to work effectively, it will be essential to have convergence of the different systems and procedures. This will involve a process of continuous improvement.

a. Whose planning system will be used?
b. How will man hours be monitored?
c. Agree cost monitoring/reporting for reports to the client and internal to the joint association 'partners'.
d. What cost code structure will be used?
e. Develop project procedures.

9 Financial Control

9.1

The joint association 'partners' must agree who pays for what. If too many costs are charged to the joint association, it could create a situation where there is no profit. You need to get into all the detail. For example: who will buy the paper clips?

a. What costs can be billed into the joint association?
b. What costs are reimbursable by the client?

9.2

How is money to be transferred out of the association? In difficult parts of the world, arrange to get money out on the basis of a unit rate for each individual sent out to work.

9.3

Who has the final say on pricing? How will tender costs be shared if the joint association does not win the contract?

9.4

How will any performance and other bond be apportioned? Avoid a total project value for performance bonds or the client will end up with more than 100 per cent, and the client will want some from both of you.

9.5

Decide on the requirements for insurance and bank accounts and so on. How will tax be handled?

10 Essentials for Success

10.1

Compatibility and commitment.

a. Mutual trust. Be equitable. Resist the temptation to score points.
b. Non-adversarial relationship.
c. Flexibility and adjustment to change. Well-developed organizational learning systems are needed.

10.2

Comparable contributions.

a. Joint problem solving
b. Convergence of systems and procedures

10.3

Shared and attainable goals

a. Win/Win strategy
b. Good planning of workload
c. Joint judgment of performance

10.4

An essential element for success between the client and the contractor is an equitable balance between risk and reward. There is no preferred form of contract, although it is most likely to involve reimbursable elements with incentives through sharing of cost savings.

10.5

Problems are:

a. No horizontal integration
b. No ability to coordinate infrastructure
c. No domestic market
d. Avoiding competition with parent companies

11 Why Joint Associations Fail

a. Poor initial planning
b. Insufficient market
c. Wrong partner – conflict of interest
d. Poor structuring of the venture

e. Poor execution – poor systems leading to little or no profit
f. Changing strategic objectives
g. Lack of availability of suitable staff
h. Incompatible cultures
i. One partner gaining at the other's expense
j. Career progression uncertain
k. No organization learning – just skill substitution

Section K   Performance Appraisals

As a project manager, you are unlikely to be directly responsible for carrying out performance appraisals except, perhaps, for your secretary. Performance appraisals are the responsibility of the line manager. Nevertheless, any project organization that is serious about project management must involve the project manager in the performance appraisal of their direct reports. This is necessary in order to make the matrix structure function effectively.

1   Purpose and Preparation

1.1

The purpose of performance appraisals is to let the person being appraised know how they are getting on and what is expected of them.

1.1.1
This is important for career development, improving the individual's performance on their current assignment, and to decide on salary increases and promotion.

1.1.2
On a project, promotion may not be possible, depending on the form of contract with the client. It may require their approval.

1.2

The line manager, project manager, and the project team member being appraised all need to arrive at an objective assessment of performance.

1.3

The feedback provided is part of the motivational processes and needs to be carried out in a carefully controlled way.

1.3.1
Avoid criticism that can produce a defensive reaction and have a negative effect on achievement and relationships.

2   The Interview

2.1

Pick a neutral, relaxing location and allow adequate time. Make sure that there are no interruptions.

a. Give adequate notice of the interview.
b. Ask the person to consider specific issues beforehand.
c. Provide an agenda with the sequence of the points that will be raised and their time allocation.

2.1.1

If you are going to be appraised, assess your own performance by creating a list of every achievement made during the period under review. Similarly create a list of those areas where you have been less successful, especially those that you think will be raised during the review. Naturally you do not mention issues if they are not raised by the interviewer!

2.2

Start with a neutral subject, such as reviewing the person's job description.

a. Review last year's targets.
b. Have their training needs been carried out?
c. What difficulties have there been?
d. Consider matters outside the control of the individual.
e. How can matters be improved?
f. Avoid comparing the person with colleagues.
g. Set measurable challenging targets.
h. Avoid personality qualities.
i. Which skills and qualifications are well used and which are under-utilized?
j. What ideas does the person have for improving the work processes or how it could be better organized.
k. What has the individual done to increase the effectiveness of their group?
l. What have they done to develop the capabilities of their team members?

2.3

Towards the end of the interview, discuss the person's aspirations and career development. Create an action plan for achieving it.

2.4

Summarise the results in terms of:

a. The levels of achievement
b. Plans for performance improvement
c. The objectives and targets for the following year

2.4.1
Jointly record all points that have been agreed and disagreed.

3   Post-interview Actions

3.1

After the interview, ask yourself how well you did and how well were the objectives met? Did you modify your views of the individual concerned?

3.2

Finally make sure that the company forms are filled in almost immediately, and review them with the responsible line manager and jointly with the individual. Sign the form.

3.3

Make sure that the action plan happens.

Section L Performance Measurement and Earned Value

Most clients need to know for certain when the project will finish and how much it will cost. Unfortunately, the more certainty that is required, the more detailed the information that will be needed.

The most commonly known technique of 'milestone' reporting merely tells you where you are a particular moment in time. Milestones are targets set as motivators. If you are late arriving at this milestone, it is too late to do anything about it. Consequently, we need to take corrective action before reaching the milestone. To do this, it is important to know where one is in the project and, once more, the more detail that will be required.

Measuring costs alone merely measures what has been spent, not what has been done. Similarly, measuring progress measures what has been done, not what it cost. We therefore need to combine both of these to measure performance.

Costs and programme in the design office are directly related to performance and productivity. A decline in productivity will cause a cost overrun. It will also mean that progress is not being achieved at the planned rate.

Serious progress measurement depends on good cost and time estimates. Your historical data will provide the relative magnitude of the various elements of your project to indicate where the big money is. The following are indicative for an oil and gas project[9]; other technologies will be different:

Design/Engineering including project management services and construction management:	18%
Equipment and materials:	45%
Construction:	37%

Construction breaks down into:	
Civils and Buildings:	25%
Mechanical and Piping:	65%
Electrical and Instrumentation:	10%

1 Design/Engineering Performance

1.1

In the design office, the key item to control is the hours expended by the design disciplines. The only way to control the job-hours effectively is to relate them to specific activities, which can be measured – the deliverables, such as a drawing, specification, procedure, and so on. Thus the starting point is to make a comprehensive list of all deliverables. Unless job-hours are linked to deliverables, project control is impossible.

---

9 The practicalities for this section have been contributed by Vernon T Evenson, Project Manager.

1.1.1

The first requirement is to define the weighted values of the performance measurement baselines for each of the engineering organizational disciplines, such as process, mechanical equipment, civil/structural, piping, electrical and instruments. The weightings are determined from the budget (originally the estimated cost) values as a proportion of the total budget for all these disciplines.

1.1.2

These discipline areas are where effective progress and performance measurement play an important role in helping to monitor and control project costs. Performance measurement is needed in order to provide realistic trends for the project manager and projections of the cost at completion for the client and senior management.

1.2

The performance measurement baseline budget is broken down for each work element. The budget for the work elements is in turn broken down into the activities to be performed. It can be expressed in man hour terms by dividing the budgets by the unit hourly rate. Man hours are the units of choice since they are easily recorded through timesheets. Nevertheless, at the deliverable level, some companies do a direct translation of the budgets into points.

1.2.1

Whilst the design will be performed using CAD, any progress measurement system generated by the software must conform to the principles of the metrics for manual drawings.

1.2.2

For example, in the piping group, the work elements might be general arrangement drawings, isometrics, piping specifications, and material requisitions. The activities for isometrics could be drawing and checking, stress calculations, and pipe supports with a budget of seven to eight hours

1.2.3

The budget for a general arrangement drawing might be 120 hours translated into 120 points. An A0 layout drawing might be 180 points. The key progress targets (milestones) might be: started – 30 points; checked – 30; approved – 20; client approval (depending upon the type of contract) – 40; and issued – 60 points.

1.3

Progress is now *earned* when these key progress targets or milestones are met. This is a binary process; they are either achieved completely or not at all. There is no apportioning in between milestones.

1.3.1

Accomplishment must be measured independently from actual man hours expended and independently of calendar time passed. It is not sufficient to say the job is half done if half the hours have been spent or half the calendar time has passed. The reporting system must determine how much of the work has been physically completed by counting the unit quantities completed for each work element.

2 Procurement Performance

2.1

The procurement function must also make a comprehensive listing of their deliverables, including requisitions, purchase orders, and sub contracts. These are also weighted by budget man hours and progress earned by achievements.

2.2

Equipment and materials can be broken down and weighted from the procurement discipline budget and listed by the budget purchase order value, which can be refined as actual values become known.

2.2.1

The expediters in the procurement function then report progress against the achieved milestones of purchase order acceptance confirmed, drawings approved to proceed, materials in workshop, testing complete, and delivery as stated on the purchase order.

2.3

Bulk materials are similarly weighted by purchase order value, and the progress milestones are order acceptance confirmed, inspected, and shipped.

3 Construction Performance

3.1

The budget breakdown weightings can be derived, as before, by the money budget values. However, for construction, physical quantities are the best basis for progress measurement. Man hours just don't work anymore. For example: how many man hours are there in a million cubic meters of earth moving?

3.1.1

Use cubic meters of earth moving, cubic meters of concrete, tonnes of structural steel, and tonnes of major equipment.

3.1.2

There are a number of options for the piping. The simplest is tonnes and, as before, budgeted versus earned man hours. Another is linear meters – measured over fittings and valves. The measurement that provides the best 'feel' for what is going on is diameter inch of welds.

3.1.3

Piping installation is really about butt welds, and from the drawings, general arrangements and isometrics, it is relatively easy to count the number of diameter inches by category (one 10" butt weld = 10 diameter inches), then measure progress against that. Use milestones for placed, welded out, hydraulically tested/non-destructive testing, and accepted.

This is really another interpretation of man hours because one coded welder hour needs 0.5 fitter hours, 0.75 noncoded welder hours, 0.4 rigger hours, 1 grinder hour, and 1 helper hour. This squad will typically complete a 30 dia inch weld in an eight-hour day.

3.1.4

Buildings, however, are measured by milestones: foundations and undergrounds complete, walls and first fixing of services, roof complete, windows installed and water tight, second fixing of interior fittings, inspection punch list, and handover complete.

3.1.5

However, if a project is poorly defined or it is a development project, there will be armies of quantity surveyors, measuring quantities and unit rates!

4   Practical Performance Details

4.1

In establishing the necessary data, be aware of the syndrome 'rubbish in – rubbish out.'

4.2

With the quantity of deliverables involved, the process will be carried out using a computer. The result will be that weightings will probably be calculated to three decimal places. This may provide a warm glow of confidence that you are setting up the controls accurately, but that is actually nonsense. You would be trying to measure design progress to a few minutes of a designer's time! What you really need is confidence that the whole number is reliable.

4.3

It is not essential to bring everything to the common unit of man hours. As stated in Part IV, Section M Reporting, paragraphs 3.1 and 3.2, the units chosen to measure progress do not have to be the same right across the whole project, provided the numerator and denominator used for calculations are the same.

4.4

At the activity level, every company has its own specific way of doing things. As a consequence, this is where the process is very subjective and can be as much an art as a science.

4.4.1

For example: in paragraph 1.2.3 above, the points have been 'back-end weighted'. This is done in order to motivate earning progress, with the achievement of the key milestone that means the deliverable is usable by other disciplines. See the caution in paragraph 1.3.

4.4.2

You might want to adjust the weighting for some civil activities because of the disproportionally high material value. Because of the high material value, civils can give an over-optimistic picture of progress in the early stages of construction.

4.5

Procurement's role in placing contracts usually ends with the placing of the contract. Responsibility then passes to someone else, such as engineering for design services and construction for subcontracts. Determining who gets the responsibility is a bit of a quagmire.

4.6

Progress measurement for underground piping needs to be split into a civil item for excavation and a piping item – a vital item to keep an eye on (see Part IV, Section O Design, paragraph 3.2j).

5   Linking Deliverables to Programme

5.1

The work scope, which has been quantified, must be linked to the programme. This is done by listing the work elements with their quantities and budget hours. Then insert a bar schedule against each element, indicating the duration for the activity. Once the work element list has been prepared, with the budget and the bar chart programme, the next step is to distribute or time-phase the budget hours over the programmed duration of the work element. This distribution should reflect the manner in which the labour will be applied. It could be a straight-line distribution (an even number of hours expended over each period from start to finish), or it could be an uneven distribution reflecting some form of loading curve indicating a mobilization or build-up to a peak activity and then a decline as the activity is completed and de-staffed.

5.2

Once the hours have been distributed to each working period over the programmed activity duration, then the sum of the distribution per period can be totalled to give an overall loading curve for the organizational entity. This loading curve will generally be a bell shape, reflecting the hours to be spent for each reporting period.

5.2.1

The bell-shaped curve produced from this process is the planned hourly loading per period to execute the work in the programmed time. See Figure V.L.1. The planned hourly period expenditure divided by the number of hours to be worked per period gives the number of staff or labour levels required by organizational element, that is: the required manning levels. It is the baseline plan reflecting the scope definition, the estimate and the planned programme.

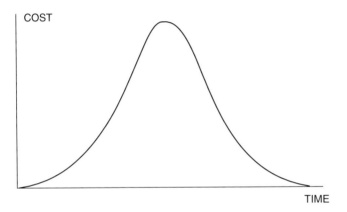

Figure V.L.1

5.3

For purposes of measurement the bell-shaped loading curve is integrated to a cumulative curve. This cumulative curve is an S-shaped curve see Figure V.L.2, (the horizontal axis is the programmed duration in months, weeks, or days, as appropriate, and the vertical axis is the cumulative hourly expenditure from zero to the total amount of the budget). The vertical axis can also be converted to a percentage scale where the total budget equals 100 per cent with percentage cumulative values for the periods in between. This cumulative S-curve showing percent complete from zero to 100 is now the baseline performance plan for the organizational entity. See also Section N 'S' Curves.

6   Recording and Comparing Data

6.1

With the project's baseline plan for performance measurement set up, the physical accomplishment at each reporting period throughout the project execution can be recorded.

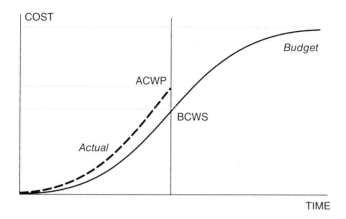

Figure V.L.2

6.2

In the Figure V.L.2 'S' curve, the project has spent more to date than planned. However, comparing actual costs to budget is not a real comparison because the actual cost may represent less or more achievement of work.

6.3

A more accurate comparison involves not only comparing the budget and actual figures but adding to the comparison a quantification of the earned value of the work done, namely, what the amount of work that has been achieved should have cost according to the planned budget.

6.4

The last step in the performance measurement process is to measure the actual expenditure incurred for the same work. This means recording the actual hours expended regardless of the physical accomplishment of the calendar time. Hours are recorded on the timesheets. The requirement is that the individual enters the code for the work element on which they have worked on their timesheet. At each reporting period, the hours charged to the work elements are summarized by work element. Consequently, the expenditures can be compared against the physical achievement and the plan.

6.5

At each reporting period, accomplishment and expenditure are compared against the plan in order to see if performance, productivity, and, hence, cost is in line with the plan and to determine if the project is on programme. Comparisons are shown in the form of progress curves where the accomplishment and expenditure are plotted against the plan. Variance from the plan shows up very clearly in this graphical form. The graphical form also allows for extrapolations to show the projected values at completion.

7   Earned Value Terminology

7.1

The time phasing of costs is best illustrated graphically.

a. Actual cost of work performed   (ACWP)
b. Budget cost of work scheduled   (BCWS)
c. Budget cost of work performed   (BCWP) – the 'earned value'

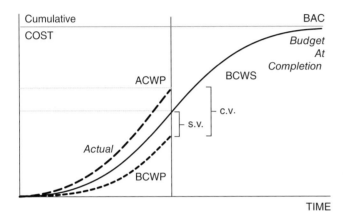

Figure V.L.3

7.2

When comparing ACWP to BCWP the common element is the work performed; consequently, the difference between them is the *cost variance* (CV).

When comparing BCWS to BCWP, the common element is the budgeted cost; consequently, the difference between them is the schedule variance (SV).

8   Useful Health Ratios or Indices

8.1

Cost performance index:
    (CPI = BCWP/ACWP [<1 represents poor performance]

8.2

Schedule performance index:
    (SPI) = BCWP/BCWS [<1 represents poor performance]

8.3

To complete performance index:
    (TCPI) = (BAC-BCWP)/(BAC-ACWP) [<1 represents good performance]

Section M   Risk and Risk List

No government should ever imagine that it can always adopt a safe course; rather it should regard all possible courses of action as risky. This is the way things are: whenever one tries to escape one danger one runs into another. Prudence consists in being able to assess the nature of a particular threat and in accepting the lesser evil.

Machiavelli, Niccolo, Translated by Bull, G., 'The Prince',
Penguin Books, 1961, p. 123.

The above quotation is a seminal statement about risk. It says that: risk does not go away. You may have put your finger on it to restrain it, but it pops up somewhere else in a different shape. This is the essence of contract strategy. How does one mould and shape the project circumstances in order to make the risks acceptable to someone else?

The project management model (see Part I, Section B Project Management Characteristics, figure I.B.1) can also be used as a risk management model. It starts with the objectives to determine which aspect will dominate the decision-making process. Risk is uncertainty that will affect the outcome of these objectives.

Next, the means to help us ensure total scope definition are the product and work breakdown structures (P&WBS – see Part IV, Section F Scope). In developing the P&WBS, it not only helps us understand and define the scope but it also identifies our areas of ignorance – the identification of a major risk in itself. The P&WBS process thus becomes our tool for total risk identification by developing a risk breakdown structure (RBS). This is done as a project team exercise, using the diverse perspectives of the different disciplines so that there is identification within the project – a major step forward in the management of risk. The team's experience and judgement is used to categorise each work package into high-, medium-, or low-risk elements; see Figure V.M.1. However, we should not forget that in this process, there is also the possibility to identify opportunities for doing something better, faster, or cheaper.

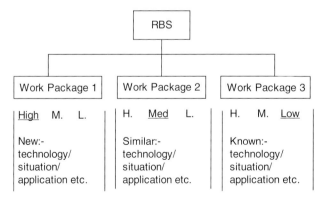

Figure V.M.1

Risk assessment falls into three main components: Identification (what can possibly go wrong?), analysis (by how much will it go wrong?), and response (what can be done to avoid it going wrong, or how can we control it if it does go wrong?).

The first risk identification process should have been performed as part of the tender or proposal preparation process. The next risk identification review should be as soon as possible after the contract has been awarded. It is an ideal team-building exercise to hold in conjunction with the client. This is not the end of the risk review process. It is a dynamic process[10] since *time* generates risk. Consequently, the review process should be continued monthly and after key milestones or other events have been achieved.

1   Process Model

1.1

In identifying risks, even with an experienced team, there is a strong possibility that they will fail to identify all the issues. The risk checklist[11] (see paragraph 3 below) acts as a memory jogger and a catalyst to the synergy of the group. It helps to generate the *'what'* of the model, the identified risk list and the risk register.

1.2

Consequently, when choosing the *'who,'* the people for the team (see Section Q Selecting and Building the Team, Subsection 1), consideration must be given to selecting people on the basis of who is best at managing risk. In choosing who to include in the group risk process, it is useful to have some 'negative thinking' people because they are more likely to identify risks.

1.3

It is probably not possible to exclude taking on major risks, such as the first of a kind, a large scale-up, technical uncertainty, or innovation. That's what project managers are for. However, don't have two of these at once – that will guarantee disaster. If you want to use new materials, use a proven application. If you want to use a new application, use proven materials.

1.4

Each identified risk must be allocated to an 'owner' in order to produce a risk responsibility matrix of *'who owns what.'* The owners remain responsible until the risk is eliminated. A risk must not be allowed to be handed over to another team member. This removes another big risk, namely, interface problems.

1.5

At the execution plan stage of the risk management model, the risk owners develop risk memos. Risk memos are response plans and options for *'how'* to handle the various

---

10 'Dynamic and Interactive Risk Management or the Project Management Risk Model', by Garth G. F. Ward presented at the Internet 12th World Conference on Project Management – Oslo, Norway, 1994.

11 Modified from a list by Perry, J. G. and Hayes, R. W., in *Chartered Mechanical Engineer*, February 1985.

'*what if*' scenarios and the relationships or impact on other risks. It may be useful to hold a workshop to find solutions to some worst-case scenarios and develop plans to avoid them. Writing risk memos (one page maximum) has a number of benefits; writing things down forces one to focus and crystallise one's plans. The risk memos should be distributed to the project team for information and reviewed by management. Most importantly, they become a library of information, which can be used by other/future projects. They record and become part of the know-how of the company.

1.5.1
Risks that are identified as high impact and severity are designated to be started at the early start date in the critical path network of the execution plan.

1.6
Use a fault tree analysis to evaluate *how* the risks should be managed as follows:

a. Recognise the risks.
b.   Examine avoidance options.
c.   Manage controllable processes.
d.   Transfer insurable risks.
e.   Quantify residual risk.
f.   Set contingencies.

1.6.1
The group process of identifying the risks, step a, will have produced an extensive identified risk list, which will need to be prioritised for effective decision-making. See subsection 2 below for the prioritising processes. The first assessment, though, may be to take no action at all about a specific risk – provided one is clear about why.

1.6.2
Avoiding options in step b are: to reduce, alter, or adjust a risk or to design it out. Alternatively, transfer the risk to someone with better skills (subcontracting) or share the risk with someone who has greater capacity for dealing with the risk (not necessarily a joint venture – the 'joint and several liabilities' may be increasing the risk).

1.6.3
Managing controllable processes in step c involves a variety of planning and monitoring disciplines: safety, quality, and execution planning, and so on.

1.6.4
If the impact of a risk is too severe, then consideration should be given to transferring the risk (step d) to an insurance company. Trying to eliminate risks in steps b and c is important since it reduces the number of risks to be considered for insurance and hence the cost of premiums. Plot severity (cost impact) against probability and use the following guidelines:

a. Low severity, regardless of probability, should be absorbed into the estimating and pricing evaluation. It is not cost effective to transfer these risks since the high probability means that the insurance premium will be high.

b. Medium severity risks should only be transferred if it is cost effective. However, the top end of the medium risk range with high probability may be uninsurable.
c. High severity risks with low probability should be insured where possible. Their possible impact on the company is just too disastrous. Their low probability means that the cost of the insurance premium is relatively low. The high probability, high risk, is uninsurable.

1.6.5
Quantifying the residual risk (step e) in financial terms is totally dependent on historical data and experience.

1.6.6
Step f is deduced from step e and emphasizes the importance of the fault tree process because it reduces the contingencies (see Section E Estimating and Contingency, sub-section 5). Consequently, it sets one's competitive edge.

1.7

The last part of the risk management model - the '*where*' are we – is to review the status of the risks at a regular risk review meeting. This is much more forward-looking and pro-active than a progress meeting that is primarily looking backwards.

1.7.1
The risks listed on the risk breakdown structure (see Figure V.M.1 above) should then move from their initial column of, say high risk, to medium risk and then to low risk and eventually disappear as they are conquered by the project.

2   Prioritising Risk

2.1

You could jump to the final stage of the prioritisation process to produce a final risk register. However, this would miss the intermediate steps that make the team really understand the issues.

2.2

The first step in prioritising the vast number of risks produced by the group identification process is to classify them (using historical data, experience, or the gut feeling of the group) for severity or impact. The impact should be assessed as the maximum likely. It can be viewed from the perspective of capital cost, schedule, and safety, as well as operability or operating costs.

2.2.1
It is unnecessary to use more than three categories for this initial assessment: high, medium, and low.

2.3

The next step is to grade the risks for probability: Certain/high probability – 3, unde-cided/medium probability – 2, and uncertain/low probability – 1.

2.4

There can be so many identified risks that it is difficult to see the wood for the trees. In this case jump to the last step and rank the prioritising process by introducing a time factor: a time scale within three months – 3, four to six months – 2, and beyond six months – 1. If these durations seem long relative to the length of the project, reduce them proportionately, but do not increase them for longer projects.

2.4.1

Select those with the highest number for: impact X probability X time. Evaluate them for: immediate action, development of contingency plans or for transfer to others, or insur-ance (see Figure V.M.2). During a risk seminar (or team building) this can be a sub-group exercise and should take the third step, in paragraph 2.5 and 2.6 below, into account.

Figure V.M.2

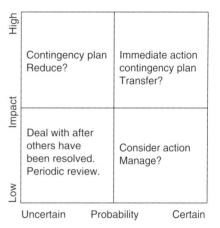

2.5

The sequential third step is to examine how difficult they would be to manage or the effort required to resolve them using: difficult to manage – 3, reasonable manageability – 2, or easy to manage – 1.

2.6

Having addressed the urgent ones in paragraphs 2.4 and 2.4.1, we can now give time to the remainder and prioritise them (see Figure V.M.3) by plotting risk (Probability X Impact/Severity), against manageability or effort required. Tackle them in the order of 1 – high risk/low effort, 2 – low risk/low effort (get them out of the way) in order to deal with; 3 those left over from 2.4 and 2.4.1 – high risk/high effort. Finally, when you have time, 4 – low risk/high manageability effort. In team building, or a risk seminar, these are tackled in subgroups before sharing/discussing the results with the whole team.

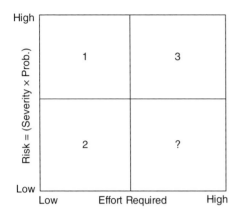

Figure V.M.3

2.7

The final step is to produce the final risk register using all four criteria and the rankings already defined of 3, 2, and 1, namely, Probability x Impact x Manageability x Time. This produces a maximum risk coefficient of 81, requiring immediate attention.

2.7.1

The risk coefficient is tabulated first on the risk register, followed by the description of the risk and then the action time factor so that it is easily seen and updated and followed by the other criteria; see Figure V.M.4. One's views of probability, impact, and manageability may change as work on the project progresses. As indicated above, time generates risk. They may need to be adjusted in the register, but time should be changed and the risk coefficient updated before each risk review meeting.

		RISK REGISTER - Week No. ...... Date.							
Risk coefficient	Description of risk		Action time	Probability	Impact	Manageability	Previous	Risk	Coefficients
27	Unknown ground conditions		3	3	3	1			

Figure V.M.4

2.7.2

Repeating this process on a monthly basis means that eventually all risks rise to the top of the list (and the project manager's attention), due to the changing value of time.

3 Risk List

An attempt has been made to list the main risk groupings/categories in the order of high impact but low probability to low impact, high probability.

3.1 Political and Economic

a. War – revolution – civil disorder
b. Political uncertainty – elections due
c. Economic downturn
d. Changes in law, imposition of decrees
e. Changes in specific legislation – financial/social security
f. Corruption
g. Currency restrictions
h. Changing boardroom politics
i. Customs or export restrictions
j. Requirement to use local organization
k. Constraints on available labour
l. Demographic changes
m. The Greens and/or other minority interest groups
n. Protesters at the site gate
o. Impact of failure: on clients, on company reputation, on suppliers, or subcontractors
p. Requirements for permits – approvals
q. Planning permission

3.2 Environmental

a. Pollution – contaminated water
b. Waste treatment
c. Public enquiries
d. Regulations
e. Preservation/conservation/reinstatement
f. Minority interests
g. Demographics
h. Archaeology/artefacts.

3.3 Legal/Contractual

a. Direct liability
b. Liability to others
c. Unlimited liability – *Never, ever, agree to this!*
d. Local laws and codes

e. Legal differences between countries of execution
f. Conditions of contract: liquidated damages, changes to accepted risks
g. Failure to sign contract
h. Insufficient tendering time
i. Joint venture

3.4   Financial

a. Interest rate changes/fluctuations
b. Inflation rate change
c. Exchange rate fluctuation
d. Availability of foreign exchange
e. Availability of client funds
f. Cash flow of client
g. Repatriation of funds
h. Loss due to default – contractor/sub/supplier
i. Contractor cash flow – late payments
j. Adequate payment for variations
k. Local and national taxes
l. Credit worthiness of contractor
m. Cost of legal decision
n. Insufficient insurance
o. Tender validity extension
p. Tender bonds unfairly called
q. Fixed price overruns – man hours/equipment/bulk materials
r. Contract budget significantly different from tender figure

3.5   Estimating/Costs

a. Budget cut significantly from original estimate
b. Estimate out of date due to late project start
c. Insufficient basic data or detail
d. Low level of confidence in estimating data
e. Lack of confidence in elements of estimate
f. Over-optimism.

3.6   Planning and Control

a. Urgency
b. Inadequate planning: at tender stage or during project
c. Complex logic
d. Schedule significantly reduced from original targets
e. Client planning system to be used
f. Subcontractors/manufacturers to use unfamiliar planning or reporting tools
g. Unable to forecast events for long project

3.7   Project Team/Personnel

a. New project manager. No experience with project type or contract
b. New project team – many new people new to company
c. Organizational – interfaces/communication/responsibilities
d. Key personnel not managed/worked on similar project or contract before
e. Lack of project manager's or team's business experience
f. Key personnel not available
g. Personnel in new positions
h. Lack of continuity of personnel from previous projects
i. Significant training requirements
j. Availability of key personnel
k. Limited resources
l. Personnel available only part time
m. Reliance on key personnel or external expertise
n. Personnel require training
o. Lack of team spirit
p. Responsibilities not clearly defined
q. Specialists'/architects' egos and other egos!

3.8   Design/Technology

a. New/unfamiliar technology, process, or equipment
b. Insufficient development time
c. Incomplete design scope
d. Evolving scope due to technical uncertainty
e. Availability of information
f. Availability of Users for involvement in early phases
g. Innovative application
h. Level of detail required and accuracy
i. Lack of confidence in basic data/project assumptions
j. Appropriateness of specification
k. Likelihood of change
l. Interaction of design with construction
m. Non-standard details
n. Non-standardisation of suppliers
o. Level of quality control
p. Reliance on external expertise
q. New hardware or new software
r. Complex or a lot of software
s. Changes to or updating of software required
t. Unfamiliar systems, interfaces with other systems

3.9   Implementation/Fabrication/Installation/Construction

a. Construction manager not involved in the early design phase
b. Delay in possession of facilities/site

c. Failure of equipment
d. Availability of equipment and materials
e. Weather
f. Quality/availability/productivity of people, both manual staff and management
g. Full time or part time working
h. Industrial relations
i. Absenteeism – sickness
j. Suitability/availability of bulk materials
k. Supply of manufactured items
l. Quality/availability/productivity of subcontractors
m. New methods, systems, or procedures
n. Unusual safety issues
o. Late design changes
p. Failure to maintain productivity
q. Failure to fabricate/construct to specification
r. Poor workmanship
s. Poor ground conditions. Do not accept unknown ground conditions!
t. Relationship of contracting parties
u. Liaison with public authorities
v. Wastage, theft, or damage to materials
w. Errors or omissions in material take-offs or bills of quantity
x. Communications with project team
y. Delay in design information
z. Poor manufacturers' drawings
aa. Access problems/restrictions
bb. Damage during transport – storage
cc. Damage during installation/construction
dd. Traffic violations/congestion charges

3.10 Setting to Work/Commissioning/Start up

a. Failure to meet process guarantees
b. Failure to meet schedule
c. Inadequate mechanical completion tests/system tests
d. Not involving commissioning personnel and/or operators early
e. Test packs not started early enough
f. Availability of client user personnel

3.11 Plan for some of the worst risks, and then you will be better able to cope with a worse than the worst risk!

4 People and Risk

4.1

It can be argued that the biggest risk of all is the people not performing their work correctly or effectively. The majority of the risks listed are the fault of people failing to perform a study, an analysis, or an investigation, or failing to train staff to have adequate skills and so on.

4.2

On the whole project managers are optimists. So remember that over-optimism will lead to under-estimation of cost and time and even to the risk themselves.

4.3

Projects don't seem to go wrong because of the big risks, despite the fact that there is limited ability to control them. Perhaps this is because people give them a great deal of attention. Projects go wrong because of the multitude of high probability, low impact risks, despite the fact that we have total ability to control them as part of the design, procurement, and construction work process. Work at them more.

4.4

People tend to be risk takers when dealing with a loss and risk averse when dealing with a gain.

5 Country Risk Assessment

5.1

Ideally a country risk assessment will be specific to an enquiry for a particular prospect that the business development people have in mind. It needs to be thorough since it should form part of the documentation for the approval-to-tender request.

5.1.1

For a client, much of the information will have been assessed prior to the feasibility stage when evaluating the market for their product.

5.2

The introduction to the country report will provide general data about the country: population, economics, rate of inflation, parliamentary system, culture, religion(s), and language(s). See also Section O, Site Checks. Most of this preliminary data can be obtained from financial institutions.

5.3

Credit rating of the country. Check with the national insurers:

ECGD	– A British government organization
NCM	– A private Dutch company
COFAS	– A private French company
EXIM	– A U.S. government department

HERMES	- A private German company
SACE	- Italian joint stock company
JBIC	- Japanese public financial institution and export credit agency

5.4 Other information more focused on doing business in the country is as follows:

a. The legal system, is it: English, European, or socialist? What are the rules for establishing a joint venture? What are the formalities for registering a local branch?

b. What are the direct and indirect taxes – national and local taxes and taxes on goods and services (VAT equivalent)? What personal taxation will apply to expatriates?

   A word of caution: we asked the tax people what we should do with regard to the importation of goods on a project in A*****a. We followed their advice. However, just as the project was coming to an end, we received a tax demand. We responded, "But, we followed your advice." They replied, "Yes, but we were wrong, and because you haven't paid the taxes you are subject to a 100 per cent fine as well!"

c. Details about the banking system and about the transfer of funds to and from the country. What are the rules for establishing a local bank account?

   When we came to closing our bank account on the same project as b above, we were not allowed to transfer the balance back home.

d. Methods of doing business: will it be necessary to establish a local company? Will a local partner, consultant, or agent be required? What local subcontractors are available? Are there any minimum rates of pay?

   We learnt that in one country, you never got paid the last payment. Consequently, you have to adjust your payment schedule to take this into account!

e. Goods and services: obtain a list of restricted goods. List materials and services that are available locally. What duties are payable on imports?

f. Employment of staff and labour: on what basis can expatriates work in the country? What are the employment conditions for local staff and labour? Will religious practices or restraints affect productivity or absenteeism? What is the industrial relations scene, and are there established trade unions?

Section N   'S' Curves

Plotting the work done (man hours, orders placed, money spent, or percentage progress) against time produces the 'S' curve shape. Unfortunately the indications they provide are retrospective; nevertheless, they indicate the status of the project and help to forecast where the project is going.

As well as the overall 'S' curve, individual curves for design/engineering, the purchase of materials and equipment, and construction should be plotted to give an overall picture of the status of the project

A common flaw with any progress measurement system (that needs careful monitoring) is that people will perform the easy items of work first. This is because that is how they get measured/rewarded with the progress recorded on the 'S' curve.

In the early stages of the project (when the basic controls are still being established), it is difficult to obtain the realistic data needed to create meaningful curves. Nevertheless, it is important to analyse the health of the project as soon as possible in the launch phase. Some suggested diagnoses of typical project situations are shown below.

1   Interpreting the Curves

1.1

Case 1—see Figure V.N.1: In the diagram, the man hours used and the actual progress are coincidental, but progress is way behind schedule.
   What's wrong?

1.1.1

If the man hours used and the progress are matching, then there is nothing wrong with the productivity—it's 100 per cent. So, the answer must be that not enough man hours are being expended. There are not enough people doing work in order to achieve the required progress. However, is there enough work available for more people at this early stage?

Figure V.N.1

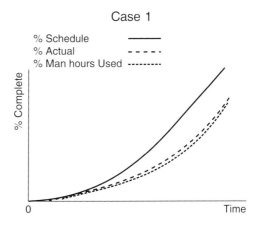

1.2

Case 2—see Figure V.N.2: In this diagram, the actual progress matches the schedule requirements, but the man hours used are excessive. Again, what's wrong?

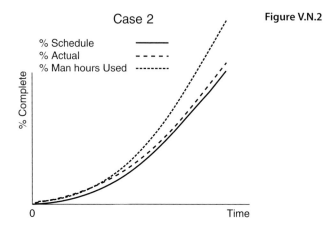

1.2.1
There is nothing wrong with the motivation of this team if they are consistently meeting the schedule. So, why are excessive man hours being expended?

1.2.2
If the motivation is good and the hours are high, there could be two or three similar reasons. Firstly, the estimate was wrong. However, if the organization is sophisticated enough to measure progress in this manner, then they are unlikely to get the estimate consistently wrong over the whole range of the project work.

1.2.3
I believe that it is more likely that the team is doing additional work that has not been recorded in the system. Is this additional work internal to the company, or is a client interfering in a reimbursable project? If this pattern existed more in the middle of the project rather than the launch phase, then I would suggest that the extra work are client changes that have not been recorded.

1.3

Case 3—see Figure V.N.3: In this diagram, the actual progress is way behind the schedule requirements, and the man hours used are far too excessive. Once again, what's wrong?

1.3.1
This project team is a mess! The project will probably never recover unless the project is stopped and started again after some serious team building. Too many people have been brought onto the project too early, and there is not enough work available for them to do.

Figure V.N.3

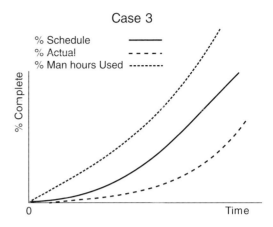

Case 3

% Schedule
% Actual
% Man hours Used

% Complete

0   Time

1.4

Case 4—see Figure V.N.4: In this diagram, the scheduled progress, actual progress, and the man hours used are all coincidental. This time there is nothing wrong. Nevertheless, the project is not perfectly under control since the cost of the man hours is over budget. The project manager has to decide whether to use experienced and more expensive people or younger people costing less than was budgeted for. However, one would require less supervision and thus save some money, and the younger group would require more supervision and cost more. If money is the only criteria, the skill is in getting the balance right. However, other factors come into play; see Section Q, subsection 1, Selecting the Team.

Figure V.N.4

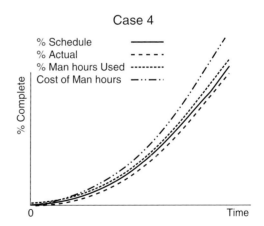

Case 4

% Schedule
% Actual
% Man hours Used
Cost of Man hours

% Complete

0   Time

1.5

A Deception?

If you can justify taking a break, say over the Christmas holidays (and in the real situation we were moving offices as well, so it seemed justified), then the 'S' curve goes horizontal as in Figure V.N.5. If you didn't start well and, despite the circumstances, you manage to get some people to carry on working (as we did), then hey, presto, you are ahead of schedule.

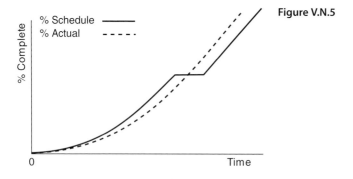

Figure V.N.5

1.6

A Self-deception
    See Part 1, Section E The Manager of Projects, paragraph 1.1, d, Figures I-E-1 and I-E-2.

1.7

Plotting the early-start (ES) dates for all activities, from the analysis of the critical path network, also produces an 'S' curve. Another similar curve is also produced by plotting the late-finish (LF) dates (see Figure V.N.6). The question now is: do you hide the LF dates before issuing a schedule of completion dates for the activities and putting the derived curve up on the conference room wall?

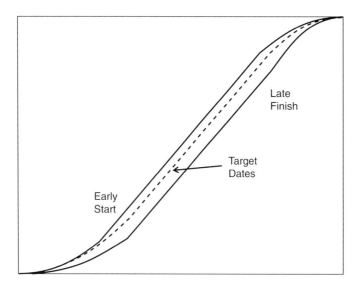

Figure V.N.6

1.7.1

At some stage, an activity will be finished late. However, since you also have the late-finish dates and associated 'S' curve, you may not be too concerned. You can see that the date achieved is within the envelope of the ES and LF curves. Consequently, you don't make too much of a fuss about it. What happens next is that more dates slip because the target provided doesn't seem to be too important. The project then slips into trouble.

1.7.2

It has happened because you showed a lack of trust in your team. Give people some flexibility and trust them; show both the ES and LF curves. Nevertheless, make it clear that if they complete after the LF date, they are in real trouble. Options are to show a target curve somewhere in between or, in order to motivate people, only issue the target curve as the visible one to aim for. Discuss these issues as part of team building.

2   Change Orders

2.1

Assuming a project of 250,000 hours and seventeen months duration, then the 'S' curve would be as shown in Figure V.N.7. This shows a *straight line* portion of 69.3 per cent to be performed in nine months, that is, 7.7 per cent per month

(a good rate of progress) equivalent to 19,250 man hours per month and, on the basis of a forty-hour week (160 hours per four-week month), 120 people.

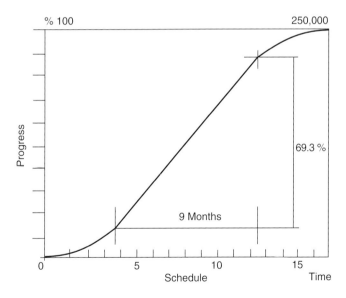

Figure V.N.7

2.2

The project has expended 75,000 man hours and is thus 30 per cent complete and on the schedule curve. However, the client has asked for a variation, a change of 50,000 man hours, making the total project hours 300,000. Consequently, the project is now only 25 per cent complete as shown by the vertical line drop in the progress curve; see Figure V.N.8. There is a problem now, since to maintain the same rate of progress at 7.7 per cent, additional people are required.

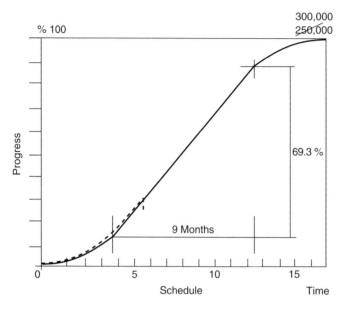

Figure V.N.8

2.2.1
7.7 per cent of 300,000 is 23,100 man hours. Therefore, 23,100 − 19,250 = 3,850 additional man hours. On the same forty hours per week basis, this equates to twenty-four additional people.

2.2.2
If additional people are not brought on board, the man hour expenditure will remain at 19,250, meaning that the rate of progress will drop to 6.4 per cent per month with a consequent two-month delay to the schedule; see Figure V.N.9.

2.3

The client will insist, nevertheless, that the project is completed to the original schedule and to 'man up' accordingly. On the basis of the man hours required to complete the project and the time available, a progress rate of 8.2 per cent seems acceptable. However, this is now becoming a high number to maintain consistently.

2.3.1
8.2 per cent of 300,000 is 24,600 hours per month. 24,600 hours at 160 per month is equivalent to (153.75) 154 people. It may be tempting to 'take the money and run', namely, to man up and get paid for the additional man hours and say that the schedule can be met because that is what the calculations indicate.

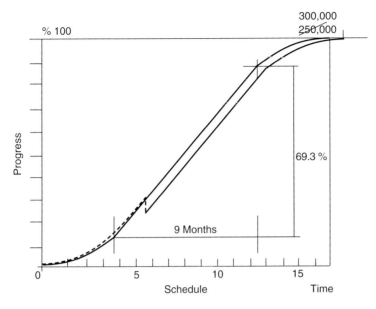

Figure V.N.9

2.4

However, we started with 120 people, and we are now proposing to man up to 154 people, an extra thirty-four people. You cannot add this number of people and expect to make immediate progress at the desired rate. A good number of the existing project team will need to stop work and explain what needs doing. The office layout of the project will need to be adjusted, there may need to be some additional team building to bring everyone on board, and finally there will be a learning curve as there would be at the beginning of a project. Thus, despite manning up, the project is still going to be late as illustrated in Figure V.N.10.

2.5

If the project is going to be late in any case, it might be better to consider not adding the extra people but working overtime. On the basis of one hour an evening or five hours at the weekend, a forty-five-hour week, the number of personnel drops to 137, that is the additional personnel drops by half to 17. Working evenings and weekends (a fifty-hour week) means that the total number required on the project is only 123 people. Three more than the original team, and theoretically, the project will finish on time. However, productivity drops off with regular scheduled overtime (see Part IV, Section Q Installation and Construction, paragraph 4.7).

2.5.1

The other option to consider is to take the extra work and to set up the thirty-four additional people as a separate project team. You might even be able to generate some competitive spirit between the two teams, based on productivity levels.

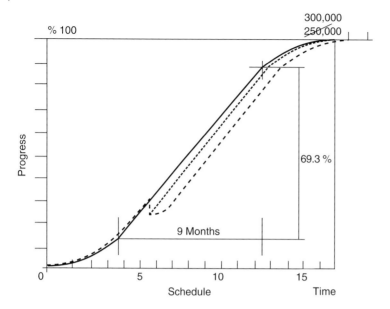

Figure V.N.10

2.6

When many individual changes have become *significant,* they should be summated and treated as a single change (as in paragraphs 2.2 to 2.4 above). Their cumulative effect should then be determined by adding a re-launch learning curve (in 2.4 above). The re-launch learning curve should be justified by being based on the historical data obtained in the launch phase. The re-launch curve should be plotted in direct proportion to the hours used and time taken to achieve the initial straight line section of the 'S' curve. You may have difficulty persuading the client to accept this reality check! See Part III, Section F Contracts, paragraph 3.7 to 3.9 about restraining the client's changes.

Section O   Site Checks

Check the country risk assessment report (see Section M Risk List, subsection 5) prepared at the tendering stage for some of the basic data. Are the data and assessment still valid?

1   Country Data

a. Available statistics
b. Geography
c. Economy
d. Culture
e. Home country embassy
f. Airport(s)
g. Official holidays
h. Banking
i. Tax system
j. Legal system/labour laws
k. Cost of living
l. Labour history
m. Construction practices

2   Site Data

a. Location/satellite map/street view
b. Local weather reports
c. Site topography
d. Existing structures
e. Transport links – road/rail/water
f. Utility supplies and costs
g. Drainage

3   Local Authorities

a. Customs
b. Tax office
c. Health services, local hospital
d. Trade unions
e. Labour agreement
f. Licencing and codes of practice

4   Suppliers and Local Contractors

a. Review number by discipline/category
b. Prefabrication/preassembly of units

c. Ownership and home office
d. Capabilities
e. Financial and commercial background
f. Suppliers for consumables
g. Fuels and lubricants
h. Availability of used construction equipment
i. Resale of surplus materials, tools, and equipment
j. Rental market

5   Labour Availability

a. Skilled labour rates.
b. Unskilled labour rates.
c. Productivity.
d. Permits required.
e. Holiday entitlements.
f. Recruitment practices.
g. Training requirement.

6   Non-manual Employees

a. Non-manual local employees
b. Local office staff
c. Regulations for expatriates
d. Qualifications
e. Recruitment

7   Housing and Camp

a. Hotels: number and classifications
b. Expatriates camp – married quarters/bachelor housing
c. Labour camp
d. Existing buildings
e. Utilities
f. Roads/fencing/drainage
g. Family living facilities and costs
h. Mess hall and social club
i. Food suppliers
j. Medical and dental facilities
k. Transport to site for staff/manual labour

8   Shipping and Handling

a. Port facilities and charges
b. Import licences
c. Custom duties
d. Customs inspection procedures
e. Time for customs clearance
f. Consider use of local agent
g. Consider bonding site
h. Sea freight rates
i. Air freight
j. Road haulage costs
k. Rail freight costs
l. Accessibility for pre-assembled units
m. Warehousing
n. Handling of materials on site

Section P   Surety Bonds

Surety bonds (provided by an insurance company) are referred to just as bonds in the project management world. Alternately, surety can be provided by bank guaranties. Their purpose is to provide the client with a sum of money in the event that the contractor defaults. The alternative to a bond is for the client to ask for a parent company guarantee.

Bonds and guarantees are so dangerous that this section is limited to the basic information that a project manager needs to know. An expert should be consulted for anything else.

It is safest to think of them as big banknotes that can be cashed by anyone!

There are three parties involved in bonds and guarantees, and each has a different interest:

The *beneficiary* (the client) wants to receive a compensatory sum of money if the contractor fails to meet their obligations

The *principal* (the contractor) does not want to pay if they have met their obligations.

The *guarantor* (or surety) wants to meet their commitment without becoming involved in possible disputes between the beneficiary and the principal concerning correct performance.

In the United Kingdom, the use of financial guarantees issued through banks has traditionally been fairly limited. The reason is, firstly, the financial status of a company can be reasonably well established. Secondly, it has been considered sufficient to rely on the contractual remedies that exist in the contract and trust in the legal system.

A common feature of guarantees is a requirement that the guarantor and contractor be jointly and severally bound under the guarantee. This gives the client the ability to make a claim against either the guarantor or contractor, although in practice it would invariably be the guarantor against whom the claim would be made.

The terms *on demand* and *conditional* are frequently used in relation to bonds. However, it should be recognised that most bonds are on demand if the terms of the bond have been complied with.

In the building and civil engineering industries, conditional bond formats have been the norm. However, in the mechanical and petrochemical engineering industries, on-demand bonds tend to be used.

The current Uniform Rules for Contract Bonds were issued by the International Chamber of Commerce (ICC) in 2000.

1   Types of Bonds

a. Tender bond
b. Performance bond
c. Repayment bond
d. Customs bond
e. Subcontract bond
f. Retention bond
g. Warranty bond
h. Maintenance bond

1.1

The tender bond is requested with an enquiry in order to coerce the contractor to complete negotiating and signing the contract, should the tenderer be chosen as the successful contractor. If they don't sign the contract, then the client has a sum of money to cover the cost of re-tendering the project.

1.2

The successful contractor will almost invariably be asked to convert their tender bond into a performance bond. The contractor basically has no option but to do so. The performance bond enables the client to use the value of the bond to get the project completed by someone else if the contractor fails to perform according to the terms of the contract.

1.2.1

See also Section J Joint Associations, paragraph 9.4.

1.3

A repayment bond is to enable the client to get their money back if an advance payment is to be made.

1.4

The customs bond makes sure that the contractor pays the customs duties that accumulate on the importation of goods.

1.5

The subcontract bond is insurance against the contractor not paying their subcontractors so that the client can use the monies to do so themselves.

1.6

The retention bond is slightly different in that it can be requested by the contractor. If the contractor would rather have the cash than let the client keep the retention monies, then the contractor can provide a bond to cover the sum involved. Nevertheless, the client is unlikely to agree to this mechanism until practical completion has been achieved.

1.7

A maintenance bond is similar to a performance bond in that it is insurance that the contractor will provide maintenance facilities and upkeep for a defined period of time.

1.8

Top up bonds exist! However, you have to have rocks in your head if you ever contemplate them.

The '*top-up*' refers to a contract condition, which is introduced by some clients. The condition provides that in the event of a call being made on a bond, the value of the bond must be made up to its original value. This is an extremely onerous condition, as in theory the client can keep on making calls under the bond with the guarantor having to 'top up' the bond each time to its original value.

2   Characteristics of Bonds

2.1

It is important to understand that the bond has a life of its own and exists as an entirely separate contract. This applies regardless of what it says in your contract and as to how the contract document may be collated and bound together.

2.1.1
The bond will be issued under the legal system of the owner's country.

2.2

Bonds have different periods of validity in different countries, and in some countries they never die. Nevertheless, it is good practice to quote a validity period since, in a dispute, it might help influence a decision.

2.3

The document should be returned to the bank as soon as possible after satisfying the contractual requirements. Bonds cost money every month that they are not returned to the bank that issued them. Consequently, it is essential that they are recovered. Help the client to find them if that is what it takes to get them back.

I spent four weeks helping an Egyptian client with their filing system in order to recover a $50m banknote.

Section Q   Selecting and Building the Team

"A round man cannot be expected to fit into a square hole right away. He must have time to modify his shape."
> Mark Twain (Samuel Langhorne Clemens) *More Trumps Abroad*, 1897.

I trust that it is self-evident that one person is not a team. With two people in a project/work environment, it is likely that job titles, rank, or years of service will skew the relationship. With three people, there will always be the problem of two people against one. It is not until there are four people that the true dynamics of a team come into play. With four people, it is also possible to have all of Belbin's original eight team roles (see the following Section R) in existence within the group – if we take into account their secondary as well as their primary roles.

As is stated in Part IV Section D (paragraphs 1.1 to 1.5) and reiterated in Sections H, K, O, P, Q, and S of Part IV, get the right people. This is one of the most, if not the most, important thing you can do. In the same way that the client project manager is relying on you to make a success of the project, so you are only as good as the people who report to you. You are dependent upon your team. As Charlie Croaker (Michael Caine) says in the film *The Italian Job*, "This is a very difficult job, and the only way to get it done is that we all work together as a team, and that means you do everything I say!" You, the project manager, then have to provide the leadership and guidance to allow the team to deliver safely, on schedule, and at the lowest possible cost.

Ideally you may be able to select from a ready-made team of people already working on a project that is coming to an end. However, you are more likely to have people allocated to you. In these circumstance,s team building will be even more important – see subsection 2.0 below.

As stated in Part VI, Section A, communication is the biggest problem in organizations. Projects are no different. Communication is a function of relationships; consequently, developing these relationships is essential. This is done through team building. You may be fortunate and be able to take your whole team away on an outdoor structured development programme of two or three days. This is not usually available, but you must do something if only in the canteen or training facility; if your company is far sighted enough to recognise the benefits of these facilities. At the very least, you can use a conference room.

1   Selecting the Team

1.1

If you are not able to select an existing team but have some choice (for example putting a team together for a proposal), then you will be spending many hours pouring over curriculum vitae (CVs).

1.2

Be wary of badly written CVs that are lists of jobs the individual has done or projects they have worked on. You want to see what they achieved as an individual or what was special about the projects.

1.3

Reject the CVs outright if the years of work experience divided by the number of projects is less than nine months. These people will never have had to live with the consequences of their decisions. They probably never started anything or finished anything. Good people will have worked on relatively few projects.

1.4

If you are fortunate, you will be able to lead the proposal effort for the project for which you will be the project manager (assuming you are successful). Doing the proposal with some of the proposed team members is an excellent team building exercise in its own right. You have an opportunity to see how people work together under pressure. You will also have an opportunity to see where there might be problems.

1.5

Talk to two or three trusted colleagues and sound them out on their views about an individual as well as their functional/line manager, not forgetting their most recent project manager.

1.6

Review their performance appraisal with the project manager who contributed to it.

1.7

Filter potential candidates by plotting qualifications (for the job/position/role) against experience (in the job/function/role). See Figure V.Q.1.

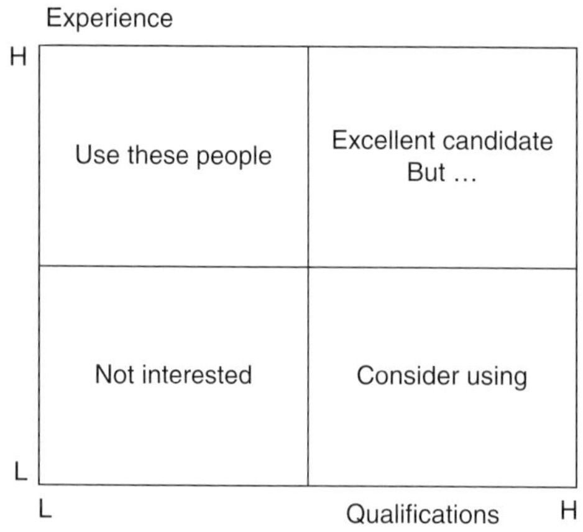

Figure V.Q.1

a. It should be obvious that you are not interested in the low – low candidates, and they should be rejected.
b. The highly qualified (expensive?) candidate but lacking experience should also be rejected. Let another project train them. Nevertheless, consider using them if they fall into the category of the next paragraph c.
c. Use some younger personnel. They will be highly motivated to get the right experience. As a bonus, their hourly rate will be less than budgeted. However, this will have to be balanced with checking of their work more.
d. The high – high person may look like an excellent candidate, but they will have a hidden agenda. They may play politics, using your project as a springboard to advance their career objectives and will leave when it suits them.
e. Your best candidate will be the one who is high in experience but is lacking in the qualifications category. For example, they have done all of the sub-functions of the role but have never been 'manager of' the function. This person will work loyally to achieve project success. They want to put that title of manager or supervisor on their CV.

1.8

Paragraph 1.7e above provides an important clue to selecting team members find out something about their hidden agenda (everyone has one) that can be satisfied by working on your project.

1.9

Weight the characteristics required for the project for each position and score the CVs against the required profile. See the following example:

	Characteristics/Score	1	2	3	4	5	6	7	8	9	10
1	General capability/flexibility									•	
2	Competence in function									•	
3	Facility/technology experience										•
4	Experience in location					•					
5	Adaptability and stability								•		
6	People consciousness							•			
7	Cooperativeness							•			
8	Cultural awareness					•					
9	Language ability			•							

a. The paper[12] from which this concept is taken makes the point that 'it is equally important to evaluate the capability of the resulting team.' Further, 'the importance of particular elements of capability for the particular project should be taken into account.'

---

12 Paper by H. McCamish (vice president, Bechtel Great Britain Limited) titled 'Team Formation for International Projects,' for the 7th Internet World Congress – Project Management Tools and Visions, 1982.

    b. For a home country project, it will probably not matter if characteristics 8 and 9 are low. However, for a project in a foreign location, you will need the number two in the discipline to be high in these areas to provide a balance of the characteristics needed. Similarly, some job functions may not require a high score for a particular characteristic. A procurement manager, for example, can have a low score with experience of the facility but would need a higher weighting for experience in the location.

    c. Do not forget to evaluate yourself. As project manager, you might be low on your knowledge of the technology. In this case you will need to make sure that there is someone in the team whose capability you trust and respect to compensate for your own deficiency.

    d. Do not be misled into thinking that a job title means competence.

    e. Have a replacement strategy. So, keep some team candidates in reserve. You will need new people or replacements at some stage during the project.

1.10

Try selecting people with totally different backgrounds. A directly applicable example, for a role involving lifting heavy equipment, would be to use someone with a background in the royal engineers. On one-of-a-kind projects, you will be able to choose people, who want a career challenge, to work on a once-in-a-lifetime unique project. Developing the (new, not done before) systems for Eurotunnel did just this, using people with a whole mixture of different backgrounds.

1.11

Finally, look at people's hobbies, outside interests, and extra-curricular activities. You may find some hidden gems of expertise.

1.12

Remember, you will have to be tough in making sure that you have chosen the right people:

> A prince [project manager] must want to have a reputation for compassion rather than for cruelty: nonetheless, he must be careful that he does not make bad use of compassion.
> A prince must not worry if he incurs reproach for his cruelty so long as he keeps his team united and loyal. By making an example or two he will prove more compassionate than those who, being too compassionate, allow disorders. These disorders nearly always harm the whole community, whereas executions ordered by a prince only affect individuals.[13]

2   Building the Team

Team building is necessary because:

---

13 Machiavelli, Niccolo, Translated by Bull, G., *The Prince*, Penguin Books, 1961.

- Team members are unaware of how their contributions can affect the project.
- The concept of teamwork is not well understood and may conflict with the culture of the organization and the experience of individuals.
- Team members tend to cover up and solve their own problems – they resist systems, which expose problems.

Having done team building on live projects[14] and having been involved in team building over a ten-year period with students at the start of an MSc project management course, the following is my perspective of the key processes involved.

2.1

Team building is concerned with moving a diffident group of people through Tuckman's development cycle[15] (see Figure V.Q.2) in order to produce a performing team with aligned objectives.

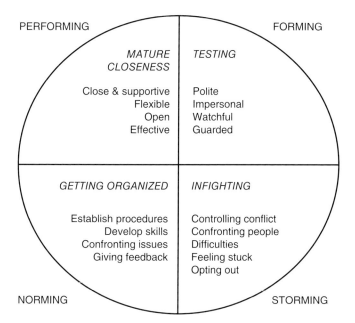

PERFORMING   FORMING

MATURE CLOSENESS   *TESTING*

Close & supportive   Polite
Flexible   Impersonal
Open   Watchful
Effective   Guarded

GETTING ORGANIZED   *INFIGHTING*

Establish procedures   Controlling conflict
Develop skills   Confronting people
Confronting issues   Difficulties
Giving feedback   Feeling stuck
   Opting out

NORMING   STORMING

Figure V.Q.2

2.2

Your training manager should be familiar with this model and work with them to develop a programme. They should be delighted that someone wants to use their expertise to the full.

---

14 See case study in John Adair's book *Effective Teambuilding,* ISBN 0-566-02605-8. Published by Gower 1986. 'Part Two Building and Maintaining High Performance Teams, page 144.
15 The 'Group Development Cycle' was first proposed by Bruce Tuckman in 1965.

2.3

Your role, as project manager, is to lead the whole process. The training manager is there to act as a facilitator and to protect you from any 'damage'. You should also agree between you when it is appropriate to let other project personnel take the lead, for example, the project controls manager leading the development of the critical path network.

2.4

The group will not become a team until it has moved through Tuckman's storming phase. This is where it is essential to use the expertise of the training manager. Getting through the storming phase can take years under normal circumstances. The challenge is how to do this reasonably quickly and in a controlled manner.

2.4.1
The development of the critical path network is one real project task that can generate a lot of emotion and is a good storming process.

2.5

Hold team building at a stage when the group is manageable. However, you need to avoid an 'us' and 'them' attitude. There will, therefore, be a need for a follow-up session(s) later so as to involve latecomers to the team.

2.6

Team building should preferably be done off site and ideally in an outdoor development centre. An exercise involving constructing a raft to cross a river (preferably in the rain!) can bring out all the emotions and characteristics needed in team building. Other smaller physical problem-solving exercises are useful to let a number of team supervisors experience different leadership styles.

2.6.1
Team building can take different forms. I once organized a coach trip, for non-technical administrative staff to take them to a site in the United Kingdom. The idea was to help them understand that a six-inch pipe was not their perceived small item but, with flanges and valves, was quite a chunky piece of metal.

2.6.2
As a bare minimum, a programme should be developed for use in an office conference room. There are good exercises involving design and construct projects with Lego bricks.

2.7

The process of team building is best illustrated by the Johari Window model[16] (see Figure V.Q.3). The model should be viewed from both an individual perspective and a group perspective.

––––––––
16 The Johari Window model was devised by American psychologists Joseph Luft and Harry Ingham in 1955.

Figure V.Q.3

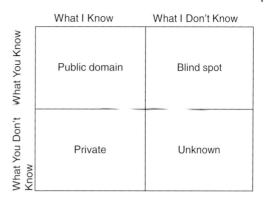

2.8

In Tuckman's forming phase, what is known about each other in the group is fairly limited, and there is reluctance for people to reveal anything about themselves. Similarly, they are unlikely to give you feedback on how you impact on them.

2.9

Exercises are needed that help the people reveal things about themselves and give and receive feedback illustrated in Figure V.Q.4. Revealing thoughts, attitudes, and personal information are important if the horizontal barrier between what is known within the group (the public domain) is to be moved downwards and so enlarge the knowledge of the group.

Figure V.Q.4

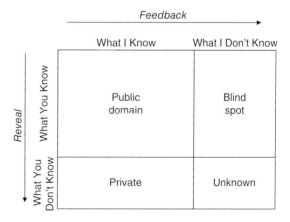

2.10

Similarly, getting feedback from people will move the vertical barrier to the right in order to reduce your blind spots and, again, enlarge the knowledge in the public domain about the group.

a. An essential revealing and feedback exercise is to complete a Belbin questionnaire (see the next Section R) and share the results with the group/team (see Figure V.R.1).

    i. Plot the team's primary role (their highest score) and their secondary role (next highest score) onto the various segments of the Figure V.R.1 diagram on a flip chart. This process is a low-risk part of the revealing process mentioned above.

    ii. Discuss and decide who might have to modify their behaviour in order to achieve a more balanced team. If you have too many characteristics that are the same (particularly shapers) in a group, the group will never achieve anything.

b. Another very effective revealing and feedback exercise is to get the individual members of the group to draw (on individual flip chart sheets) how they see themselves and their role within the team. A variant of this is how they view the project or how they see the project in the context of the company. People will draw pictures of issues that they would not otherwise reveal.

    This exercise is best carried out after dinner and a visit to the bar. Castles, bow and arrows, and daggers in the back are very interesting to have explained.

c. Get the training manager to come up with a variety of exercises involving creativity, problem-solving and different management styles. Let different supervisors lead different exercises.

2.11

There are a number of topics that should form part of the team-building agenda that are real work project issues that need to be resolved. For example:

a. Developing a project motivating phrase
b. Agreeing the communication mechanisms
c. Developing the product and work breakdown structure
d. Developing the proposal risk analysis
e. Planning the project and developing the network
f. Producing a mission statement for the project
g. Review of lessons learned from previous projects
h. Improving execution methods using VM and VE. See Section S.

3 New to the Team

3.1

If you are a project manager who is new to an established team, or if there is a new senior member of the team, a different approach is needed. Hold a series of one-to-one interviews with the direct reports. If you are the new project manager, you then need to make a presentation to the whole team on yourself and your expectations.

Section R   Team Roles

Understanding people; The graduate with a Science degree asks, "Why does it work?" The graduate with an Engineering degree asks, "How does it work?" The graduate with an Accounting degree asks, "How much will it cost?" The graduate with a Liberal Arts degree asks, "Do you want fries with that?"

Source Unknown

Dr Meredith Belbin's self-perception inventory questionnaire,[17] answer grid, and descriptions of the team roles are available on the Internet. Additional analytical services have to be purchased. For a project, I prefer the original eight team role questionnaire. The nine team role questionnaire contains the extra role of a specialist. They are in effect an extreme plant (see Paragraph 1.3 below). As a project manager, you probably do not need a questionnaire to identify these people. Further, you will recognise that they are not really team players. They make their expert contribution and then opt out.

Belbin's questionnaire is about behaviour, and whilst it has some elements of fixed personality traits built into it, behaviour can be changed. Consequently, it should not be held on a personnel file. In fact you will want to make use of a person's ability to change their behaviour/team role and work at exhibiting the characteristics of another role (see paragraph 2.10, a., ii, in the previous Section Q).

The highest score from completion of the questionnaire is the person's primary role, and the next highest is their secondary role. Sometimes people have two or three secondary roles. Younger, less-experienced people can exhibit many roles since they have not yet become set in their ways. Conversely, older people, who have become set in their ways will show fewer roles. However, the intelligent, experienced person will realise that different behaviour is needed in different situations and will, consequently, exhibit a variety of roles.

1   Specification of the Eight Team Roles

The title of nearly all the roles is very descriptive of what the role is all about.

1.1   Chairperson/Co-ordinator

I prefer the original title of chairperson. Chairing is different from coordinating. As you would expect, the chairperson sets priorities and organizes and coordinates the group. They also control the activities and contributions of the group. This involves the clarification of problems and group objectives, assigning tasks and responsibilities, and encouraging group members to get involved in achieving objectives and goals. They should be calm, self-confident, and controlled.

---

17 R. M. Belgin, *Management Teams: Why They Succeed or Fail*, Heinemann UK: London, 1981.

1.1.1   Positive Qualities:
They bring out the best in others by treating and welcoming all potential contributors on their merits and without prejudice. They have a strong sense of objectives.

1.1.2
Allowable Weaknesses:
   They are of moderate intellect and can lack creativity.

1.1.3
Comments on Belbin's 'Positive Qualities and Allowable Weaknesses':
   You might be upset at being 'moderate or no more than ordinary.' What Belbin is getting at here, is that you do not necessarily have to be cleverer than the group in order to manage it. However, you do need to be at the same level as the group. If the group is smarter than you are, then they will get away from you, and you will lose control of the group.

1.2   Shaper

The shaper wants to get on with the job. They are a driving force that challenges, argues, and disagrees. They unite the team's effort and push them forward to decisions and actions. They are achievement motivated, extroverted, impatient, have a low frustration threshold. They are competitive and respond to challenge. They have good insight. They are a non-chairman leader. They are highly strung, outgoing, and dynamic.

1.2.1
Positive Qualities:
   They bring a sense of purpose to the team with drive and a readiness to challenge inactivity, ineffectiveness, complacency, or self-deception.

1.2.2
Allowable Weaknesses:
   They are impulsive and prone to provocation, irritation, and impatience.

1.2.3
Comments on Belbin's 'Positive Qualities and Allowable Weaknesses':
   There is little to say here; that's what shapers are like. However, the strong shaper needs to work at not showing the 'allowable weaknesses' traits if they want to keep the respect of their team.

1.3   Plant

The Plant thinks differently; see the quotation at the start of this section. They contribute original and creative ideas and strategies for achieving the objectives adopted by the group. This role brings a breadth of vision, creativity, imagination, and innovation for solving difficult problems. But they don't necessarily communicate their ideas effectively. They are individualistic, serious-minded, and unorthodox.

1.3.1
Positive Qualities:
  They are clever and have imagination, intellect, and knowledge.

1.3.2
Allowable Weaknesses:
  They can be 'up in the clouds' and inclined to disregard practical details or procedures.

1.3.3
Comments on Belbin's 'Positive Qualities and Allowable Weaknesses':
  The allowable weaknesses are very logical. Once you start to manage creativity, it dries up. You want someone who thinks 'outside the box' and stops the group from going down 'group think' tramlines. They are obviously most needed in the early stage of each project phase and good for helping with project problems.

1.4   Resource Investigator

This role keeps the team in touch with the environment outside the group by exploring opportunities and identifying ideas, information, and resources. Performance of this role involves developing a network of contacts and coordinating and negotiating with other groups and individuals. They are extroverted, enthusiastic, and curious.

1.4.1
Positive Qualities:
  They have a capacity for improvisation and contacting people and exploring anything new. They have an ability to respond to challenge.

1.4.2
Allowable Weaknesses:
  Their interest is likely to wane once they have 'cracked the task'.

1.4.3
Comments on Belbin's 'Positive Qualities and Allowable Weaknesses':
  As a project manager, you may well be low in this aspect in which case, this is an unusual example where you might want to select someone based on their Belbin profile. As a project manager, you will initiate many scenarios, and if 'you lose interest once the initial fascination has past', you will be in trouble. Hire an assistant/secretary who is a high completer finisher.

1.5   Team Worker

These people build relationships. They promote unity by creating and maintaining a team spirit. They improve communication by listening and bringing people into a discussion. They provide personal support and warmth to group members and smooth over tension and conflict. They are socially orientated and rather mild and sensitive.

1.5.1
Positive Qualities:
  They have an ability to respond to people and situations and to promote team spirit.

1.5.2
Allowable Weaknesses:
 They prefer to avoid confrontation and can be indecisive in critical situations.

1.5.3
Comments on Belbin's 'Positive Qualities and Allowable Weaknesses':
 They may be seen as indecisive when pressed to support one point of view or another in a discussion. They will be more concerned with the cohesion of the group. They will be reluctant to express a view which, in effect, means they are taking sides. They want to maintain relationships with all parts of the group.

1.6 Implementer/Company Worker

The company worker gets on and does the job. They are concerned with the practical translation and application of concepts and plans into manageable tasks. This entails a down-to-earth outlook, coupled to perseverance in the face of difficulties. They are conservative, reliable, responsible, and predictable.

1.6.1
Positive Qualities:
 They are logical, orderly, and hard-working, with practical common sense and self-discipline.

1.6.2
Allowable Weaknesses:
 They are slow to respond to change and new possibilities and can lack flexibility.

1.6.3
Comments on Belbin's 'Positive Qualities and Allowable Weaknesses':
 The positive qualities are just what you want in your discipline and project engineers.
 Whilst the allowable weaknesses sound harsh and negative, they are in fact very positive qualities for people who get on and do the work. The team has agreed on what and how the work has to be done, and you don't want to experiment with new ideas at the execution/implementation stage.

1.7 Monitor Evaluator

This role involves analysing ideas and proposals being considered by the team. They bring a dispassionate analytical logic to situations, and they prevent the team from committing to bad decisions. It is important for the monitor evaluator to point out, in a constructive manner, the weaknesses of proposals being considered. They are sober, unemotional, and prudent.

1.7.1
Positive Qualities:
 They have judgement, discretion, and are strong-minded.

1.7.2
Allowable Weaknesses:
   They can lack inspiration and may be demotivating to others.

1.7.3
Comments on Belbin's 'Positive Qualities and Allowable Weaknesses':
   You want your project controls manager or project office to be staffed with people with these characteristics. You want them to take a hard line when evaluating progress. Monitor evaluators are also useful people to stop the 'group think' process. However, this may frustrate the plant who has come up with the ideas and others who are keen to get on with the work.

1.8   Completer/Finisher

The Completer finisher dots the 'i's' and crosses the 't's.' They are systematic in ensuring that the group's efforts achieve appropriate standards and that mistakes of both commission and omission are avoided. They bring a conscientious approach to quality and standards of performance and meeting deadlines. They are painstaking in checking detail and the search for errors. They maintain a sense of urgency within the group.

1.8.1
Positive Qualities:
   They have a capacity to follow through and are thorough/meticulous.

1.8.2
Allowable Weaknesses:
   They can be anxious, with a tendency to worry about small things and can show reluctance to delegate and 'let go.'

1.8.3
Comments on Belbin's 'Positive Qualities and Allowable Weaknesses':
   A rare characteristic in project people and consequently valuable. They have just the attitude needed when deliverables have to be issued.

2   A Suggestion for a Project Manager

Project managers do not have to be shapers, but in certain industries, such as the process industry, they are most likely to be strong in this area. I have come across construction managers who are off-the-scale as shapers. I have known successful project managers who have various combinations of Belbin roles. However, being bold, my choice would be:

- Shaper – chairperson
- Team worker
- Monitor evaluator

3   Matching the Roles to the Project Process

3.1

Matching the Belbin roles in pairs, we can see that they combine together to form a useful project cycle (see Figure V.R.1). Further, to some extent, the emphasis on the roles rotates clockwise as the project moves through the phases starting with the project management/leader role.

a. The chairperson and shaper provide *direction*.
b. The plant and resource investigator provide *information*.
c. The team worker and company worker *deliver and produce*.
d. The monitor evaluator and completer finisher *control*.

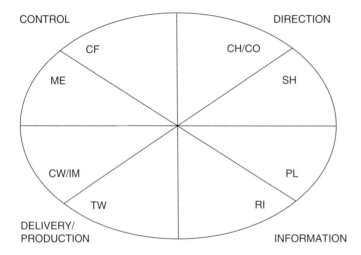

Figure V.R.1

3.2

Examine the detail of each of the eight roles, and you have an ideal combination of skills to achieve any task:

a. Setting objectives and assigning responsibilities.
b. Being an achievement-motivated leader.
c. Putting forward ideas and strategies.
d. Identifying information and resources.
e. Improving communication by providing personal support.
 f. Practical translation and application of concepts and plans.
g. Evaluating feasibility for achieving objectives.
h. Achieving standards and maintaining a sense of urgency.

Section S   Value Management/Engineering

Value analysis (VA) tends to focus on the end product, Value engineering (VE) tries to improve the design before it is incorporated into the next stage, and Value management (VM) tries to take a holistic approach. VA tends to be done by an individual, and VE and VM tend to be carried out using a quality circle approach. Quality circles are a multidisciplinary group brainstorming process for identifying and solving quality problems. They all try to do much the same thing and are traditionally most effective in the mass-production manufacturing environment. Nevertheless, there is still some benefit to be obtained in the one off project environment since the work processes are repetitive in this context.

The main emphasis of VM/VE should be in the conceptual phase whilst the owner is formulating the project. However, it will depend on the contracting arrangements for the contractor, and the most effective area is likely to be during the feasibility study. It is about identifying the functional requirements of a project to achieve an optimum solution for the least cost.

Value is regarded as a measure of how well an organization, project, product, or service satisfies stakeholders and achieves the objectives in relation to the resources consumed. Nevertheless, value management should be regarded as a means of finding a better way of doing things rather than just a cost-cutting exercise. It is a creative, innovative, group process.

During the early development stages of a project, the objective of value management is to produce a technical scheme that will produce an attractive return for the investors on their capital investment. The focus will be on life-cycle costing. In commercial building development, the focus is on the use of space and functionality. It is more business focused.

During the later execution stages, it becomes more specialist- and engineering-focused where the objective is to optimise the plant layout for piping, cabling, and steelwork. Alternately, in building development, it will seek to optimise the building services in order to reduce the use of the budget and or operating costs.

VM/VE can be applied at numerous stages as the project moves through the various phases of development. Nevertheless, the principles of value engineering stay the same. It should be identified as a key milestone(s) in the project schedule.

As a project develops, the number of people who can be involved increases considerably, from a few generalists in the early phases to a large number of specialists in the execution phases.

1   VM/VE Process

1.1

The VM/VE starts as a subjective process and eventually becomes more objective with the focus on benefits and costs. It has four distinct phases:

a. Information gathering. Do not forget lessons learned from previous projects.
b. The creative group process.

    c. The analysis stage. Cost analysis, functional analysis, understanding requirements.

    d. The development of a proposal for submission to the top management team.

1.2

The process should be initiated or led by the project manager to demonstrate commitment to the process. They need to discuss the number of people to be involved in the process and the departmental representation.

1.2.1

The project manager must specify the terms of reference of the group. Is it advisory, or does the project manager request presentations on ideas where they might want to exercise a veto on a particular line of attack?

1.2.2

Alternatively, the project manager may delegate authority to the team to make changes on its own without further permission.

1.3

The project manager defines:

    a. The objectives for what is to be achieved
    b. The issues involved
    c. The evaluation criteria
    d. The process to be used
    e. The ground rules for timing, breaks, chairmanship, records, and so on
    f. The reporting relationship to line management and explains the purpose of the group to the remaining team members.

1.4

Some pre-work should be issued to be done as individuals or in pairs.

1.5

Team members post their pre-work ideas on the wall and ask for clarification or additional explanation. There are no bad ideas – only ones to be improved.

1.6

Issue everyone with their own sheets of flipchart paper and wall space for them to write up their own ideas. Brainstorming rules apply where no idea is ruled out. See Part VI, Section K Problem Solving Techniques, subsection 1.

1.7

Every idea must have a cost saving assigned. The cost engineer, therefore, gives a quick 'guestimate' (quick rather than accurate) of the cost saving for each idea.

1.7.1
Quality assurance should be involved to ensure that the technical integrity of standards is not compromised.

1.8

The team agrees priorities for discussion (see Part VI, Section H Prioritising Techniques) and the interaction between ideas. Choices have to be made:

a. Price or purchasing cost
b. Quality or reliability
c. Materials or design
d. Production or manufacture

1.8.1
Do not make changes to two areas at the same time. For example: do not change the material and the design. See Section M Risk and Risk List, paragraph 1.3.

1.8.2
The group discussion bounces around to find new ideas. The group leader for this session allows the debate, discussion, to continue for as long as the creativity is effective. They write the ideas up on the wall as they occur and the cost engineer evaluates the cost savings.

a. Evaluate:
    i. Substitutions
    ii. New materials
    iii. Alternative processes
    iv. Improvements
b. Modify:
    i. Adapt parts
    ii. Shape/size/colour
c. Change:
    i. Non-standard elements
    ii. Manufacturer
    iii. Suppliers
d. Examine Procedures:
    i. Change working methods
    ii. Evaluate necessity of each step

1.9

The team selects (see Part VI, Section H Prioritising Techniques) the best set of ideas for development. They decide on the sequence for the discussion and the time allocation for each idea. Team members then retreat to work areas and write down an approach to be pursued for each idea. The outline plans developed include evaluations for:

a. Risk
b. Safety
c. Feasibility
d. Acceptability
e. Quality
f. Cost-saving efficiencies – defined by how much it will cost to make how much saving.

1.10

A final prioritised list is then developed within the delegated authority given to the team.

1.11

The team members make presentations to the top project management team (the project manager and their direct reports) with the preliminary implementation plans, indicating the savings made and the associated risks and liabilities.

1.11.1

Alternately, if the top management team or client is directly involved in the process, the presentation will be made to the client or corporate management.

1.12

The top management team (or client) undertakes to review the individual plans and respond.

1.13

The project plan is modified to incorporate the agreed changes and implemented.

a. Monitor progress
b. Quantify savings and improvements
c. Identify elements for re-examination

2   Group Process

2.1

The VM/VE process is totally dependent on a good group process. Thus if client, joint venture partner, or consortium partners are to be involved, then it becomes an extremely difficult process. People become too focused on defending their own interests.

2.1.1

Do not underestimate the difficulty of creating a group that will work together creatively and thinking outside their normal work constraints and company mindset.

2.2

It requires Edward de Bono's *Six Thinking Hats*[18] creative processes.

2.3

It will be most effective if the work is done in an offsite workshop environment.

2.4

The team should be comprised of the people who do the work (a supervisor may well need to be involved at some appropriate point). Also, a cost engineer estimator who has a good knowledge of cost algorithms and can generate numbers quickly is necessary.

2.5

The process will require achieving consensus among a group of specialists with divergent views. Consequently, consensus with qualification will be required – "I don't necessarily agree, but I can live with it." See Part VI, Section B Leadership and Motivation, paragraph 1.2.

---

18 Edward De Bono, *Six Thinking Hats,* Penguin Books, 1990 & 2016. ISBN 9780140296662. Also http://www.debonogroup.com/six_thinking_hats.php

PART VI

Skills Check Lists

Section A Communications

"Then you should say what you mean," the March Hare went on. "I do," Alice hastily replied; "at least – at least I mean what I say – that's the same thing, you know." "Not the same thing a bit!" said the Hatter. "Why you might just as well say that 'I see what I eat' is the same thing as 'I eat what I see'!"
Alice's Adventures in Wonderland by Lewis Carroll (Charles Lutwidge Dodgson).

"When I use a word," Humpty Dumpty said in a rather scornful tone, "it means just what I choose it to mean – neither more nor less."
"The question is," said Alice, "whether you can make words mean so many different things."
Through the Looking Glass by Lewis Carroll.

Surveys have shown that the biggest problem in organizations is communication. This is because of the variety of communication channels and because they are rarely planned.

The objective of communication in business is: 'to transmit information (however complex) accurately and concisely from one person to another in the most easily understood way'.[1]

Communication can be grouped into three main categories used at different levels in an organization:

Formal	Semi-formal	Informal
Normally external	Internal	No written constitution
(contracts)	(job procedures)	(requires team building)
Takes time	Careful design	Fast to respond

Set out below (in loose groupings) is a list of communication channels and their allocated categories, together with suggestions for how they could be used on a project. Discuss these channels with the project team as part of the team-building agenda. Decide how they are to be used and then specify them in the coordination procedure (see Part V, Section B) with clear instructions. Make people stick to the rules – particularly for topics such as e-mails.

Only 7 per cent of a message is received via the words. The remaining parts of a communication are received from how things are transmitted via the voice and body language. Consequently, the order of preference for communications should be:

First	Face to face, in person
Second	Face to face, video conferencing
Third	Voice, via telephone
Fourth	Written words via e-mail

Written internal communication (for example, e-mails) should be discouraged in favour of people meeting each other.

---

1 An excellent definition from a documentation and training organization, Plain Words Limited, 2012.

See also Section E Personal Skills, subsections 1 to 8 Interactions with Others.

1   Correspondence

1.1

All communications to the client should be coded, numbered consecutively, dated, and recorded in a log. The details of the coding and numbering system will be identified in the coordination procedure.

1.1.1

Procurement should also maintain similar logs for communications with suppliers and subcontractors and other links in the supply chain.

1.2   Letters

1.2.1

Letters to the client are formal communications and should be signed only by the project manager. However, you are not going to be writing all the letters. Make sure you spend some time training your team into writing the letters in the style or format that you think is appropriate. It will save you a lot of hassle later. See Part V, Section A Completed Work.

a. If you are responding to contractual or contentious issues, put your response to one side for twenty-four hours and reread and edit it before sending.
b. Have a clear subject heading and restrict the letter to that one subject. The subject should preferably be compatible with the filing system subjects.
c. The address of the person you are writing to should be on the left-hand side. Use 'Yours faithfully' for people you haven't met or spoken to. Use 'Yours sincerely' for people you know, such as your opposite project manager.
d. State the purpose of the letter in the first paragraph. The last paragraph should state what action you expect from the recipient.

1.2.2

Remember that legally, offers and acceptances for goods and services by letter are deemed to have been made when they are posted.

1.2.3

See Part V, Section B Coordination Procedure, paragraph 2.2, for the administration rules covering letters.

1.3   Memoranda

1.3.1

These are semi-formal and are often used to communicate internally to someone on a circulation list rather than the addressee, for example, your boss's boss!

1.3.2

In today's electronic world, these will be in e-mail format. Memos in paper format will be used infrequently to emphasize the importance of the subject. A record of a paper memo is also more likely to be retained.

1.3.3

Memos of two pages should be written as separate documents and attached to an e-mail, concisely stating what its purpose is.

1.4 Facsimiles

1.4.1

Faxes should be treated as letters and be formatted and recorded as such. Their only advantages are the speed of transmission and documents are transmitted without taking extra hard copies.

1.4.2

Legally, offers and acceptances for goods and service' by fax are deemed to have been made when they are received. Consequently, for any important and, most importantly, any commercial communications, the fax should end with a statement: "Please sign and date receipt of this fax and send it back to us."

This is necessary because one cannot prove that a fax has been received. The, now old-fashioned telex had an advantage that is missing with facsimiles, namely, it had an electronic 'handshake.' As a result, it could be proved that the message had been received.

E-mail and the ability to send attachments, really makes the fax redundant. However, the above comment concerning acknowledging receipt is also valid for e-mails.

1.5 Transmittal Letters

1.5.1

Transmittals are formal standard forms that act as cover letters for sending drawings, minutes of meetings, reports, and other documents to the client.

1.6 Circulation Lists

1.6.1

There is no doubt that paperwork is a real problem for most people, so do not add to it by indiscriminate circulation of documents, memos, and letters. Develop limited circulation lists for various topics, with only those names of people who need to know. You will have to explain to members of the team that you are not keeping secrets from them but endeavouring to make communication more efficient. In any case, copies are always available in the files. Get this approach agreed during the team-building process.

1.7 Post-it Notes

1.7.1

These should only be used as reminders of something that has already been more formally communicated – for use by the project manager who gets in early before the rest of the team. However, personal contact gives an opportunity to gauge the reactions of the people being reminded.

2 Documents

2.1 Coordination Procedure

2.1.1
This document (see Part V, Section B) establishes the formal communication channels for the project.

2.2 Minutes of Meetings

2.2.1
In a project context, minutes will not be a blow-by-blow transcript of the discussion but notes of the key points and the actions to be taken together with the person responsible. However, remember minutes are formal records that will be relied upon at later stages of the project and are likely to imply contractual obligations. Therefore, who writes them is important.

2.2.2
It is also useful for, say, the project controls coordinator/supervisor/manager, to write up all the mundane administrative/progress data minutes before the meeting.

2.3 Responsibility Matrix

2.3.1
The responsibility matrix (see Part IV, Section C Getting Organized, paragraph 3.1) is an essential tool formally defining who is accountable, who is consulted, and who is copied for information concerning key project functions and documents.

2.4 Notices

2.4.1
These are used formally when you want everybody to get the same message without the distortion produced by the verbal relaying of a message.

2.5 Procedures

2.5.1
As project manager it is easy to sign off the formal procedures as approved without reading them. If you do so, you may be signing away your decision-making and management control capability. So take the time to read them and question the authors.

2.6 Reports

2.6.1
Reports are mainly formal records of investigations and a record of project progress. They become a primary source of information at later stages in the project. See Section L on Report Writing.

2.7 Newsletters

2.7.1
These can be useful for formally updating the team on issues that they are not normally involved with, without the formality imposed by a presentation. They are most useful in updating personnel in remote locations or team members who are out of the office, for example, expeditors. It helps to make them feel that they have not been forgotten.

2.8 Brochures

2.8.1
These are primarily a marketing tool and require careful design. If you intend to produce a brochure, then you will need to start collecting material at the start of the project.

2.9 Documents – General

2.9.1
These are the formal definition of the project, for example, the terms of reference, the statement of requirements, the product breakdown structure, specifications or requisitions, whereas sketches are informal documents to aid the development of the formal drawings. The formal documents to be used and distributed on the project should be listed on a document distribution matrix, forming part of the coordination procedure, (See Part V, Section D for a list of project documents).

2.9.2
Controlling the issue number of documents is a key concern. Everyone must be using the same/latest information. A document control system is essential.

3 Electronic Media

You need to be aware of (and possibly conversant with) the available technologies.

The email to set up the three-way telephone call was short and to the point. One sender and three recipients – including Amy … "Amy can help us to set up time next week," said the sender. Amy quickly responded with times and dates. Three emails later, Amy had coordinated our calendars and set up the call. That would have been impressive in a human, but Amy is a computer programme. …[2]

However, you may find that the older manual mechanisms are just as, or more, effective.

3.1 Video Conferencing

3.1.1
It is a more difficult process to use than it seems. You either see the whole group, but cannot distinguish expressions or body language, or you see one or two people in close up but do not know or see the person making the interruption.

---

2 From an article "Who needs Humans?" by Charles Arthur, in BA's Business Life, October 2015.

3.1.2

An effective chairperson is essential, and rules need to be agreed. Discipline in preventing interruptions and over talk is vital. There must be an absolute bar on any side conversations between team members.

3.1.3

Make sure everyone knows each other from previous face-to face meetings. Spend a little money and get everyone to meet each other (during team-building sessions) before using this mechanism.

3.1.4

For projects with multiple locations, this can save a lot of travel costs (time and money) in getting people together to resolve problems.

3.2   Telephone

3.2.1

The problem with the telephone is that the interruption always seems to take precedence. Consequently, it can be very disruptive to getting work done.

3.2.2

As indicated in 1.1 of this section, make sure that a summary of any formal conversations are recorded and logged. All agreements made with the client on the telephone should be confirmed in writing (through the project manager) within, say, three working days.

3.2.3

With careful planning, significant monies can be saved by restricting the telephones of certain members of the team from making national or international calls.

3.3   Telephone Conferencing

3.3.1

Like video conferencing, the quality of the process is very dependent on the technology. Again it can be useful in saving the costs of people having to travel.

3.3.2

Once more, a chairperson is needed. It needs rules to be agreed and requires discipline. Interruptions/over talk can stop everyone from hearing what is being said.

3.4   Mobile Telephone

3.4.1

These cannot be ignored but should only be regarded as informal personal communication tools. Nonetheless, expeditors and inspectors in the field will find them useful for instantaneous reporting.

3.5   E-Mail

3.5.1

There is no doubt that this mechanism is the biggest problem and poses the biggest challenge. When people tell me that they have 200 to 300 or more e-mails to read every morning, I know that we have not yet learnt how to use this communication channel effectively.

3.5.2

If it must be used, which it will be, the e-mail should clearly indicate its communication category by being formatted as a formal letter or as a semi-formal memo. However, since all letters to the client must be signed by the project manager, they are less likely to be used in this manner. Consequently, they become an internal communication tool and can take the place of the hard-copy memo. However, unless hard copies are taken as a policy, there is no permanent record, and they then take the place of a verbal communication. E-mails, therefore, become the non-oral communication tool between remote locations.

3.5.3

Discuss the problems of e-mails with the project team, decide how they are to be used, issue clear instructions, and stick to the rules.

3.5.4

Here are some guidelines to help make e-mails more effective:

a. Try to make your objective clear in the subject heading, that is, what you hope to achieve with the e-mail.
b. Develop a coding system to be used in the heading to indicate what is expected of the recipient. For example, A.R. – Action Required. R.R. – Response Required. Include a date for completion of the action.
c. If you have to write a long e-mail (people are less likely to read them properly), summarize it briefly in the e-mail and put the rest of the detailed material in an attachment.
d. Break up large chunks of text with headings and bulleted lists.
e. Divide the text into important facts, actions, targets, reasons and supporting background, and justification information.
f. Also see Section L Report Writing, subsection 8 that provides some guidelines to make the text more effective.

3.6 Text Messaging

3.6.1

This mechanism should only be used for informal communication. It has the disadvantage over voice communication in that voice tones cannot be interpreted. Nevertheless, it has the advantage of being less disruptive in that the receiver of the message can read the message when it is convenient to them. Consequently, it should not be used for urgent communications – unless it is the only mechanism available.

3.7 DVDs/CDs/USB Memory Sticks

3.7.1

Used for copying presentations to remote locations and for publicity purposes.

3.7.2
If you intend to make a video of your interesting project, you will wish that you had started it at the beginning of the project.

3.7.3
CDs are useful for project long term records where hard copy is not a requirement.

3.8　Internet and Intranet

3.8.1
The Internet becomes a black hole, sucking in man hours. Think about banning it except for those personnel with a genuine need to research companies, for example, the project procurement manager.

3.8.2
The Intranet is the source of all company standard documents and data. It can also be useful to establish a project website for project documents and data, as well as using it as a problem-solving forum.

3.8.3
A file transfer protocol (FTP) site is used for transferring files that are too large for e-mail attachments and get blocked by the company IT firewall. E-mail the recipient with the link to the site and specific document.

4　Oral

4.1　Face to Face

4.1.1
Oral communication takes on the formality, or otherwise, of the situation. For example, a meeting will be formal, but a discussion with the client in a pub is semi-formal. A discussion with a member of the client team can never be informal.

4.1.2
Learn to read body language. Individual gestures can reveal how someone is feeling (people cannot disguise making certain expressions or gestures). However, it is combinations of body language that give the most certain confirmation of someone's feelings. For example, arms folded across the chest gives a negative message, but they may only be relaxing. On the other hand, arms folded and pushing their chair backwards is a definitive opting-out message.

4.2　Meetings

4.2.1
When people are in meetings, they are not working. Consequently, it is important to find mechanisms to keep them short. See Section C Managing and Conducting Meetings.

4.2.2
In projects that are spread out over a number of work sites, it is useful to have a quick (ten minutes) update or morning report meeting. This makes sure that everyone is up to speed with what is going on and what is anticipated during the day. The meeting should be held standing up so that people don't get comfortable and prolong the meeting. Alternate the location of the meeting between the different sites.

4.2.3
Meetings are used in different ways in different cultures (see Part V, Section C Cultural Issues, paragraph 3.1 m and n). For example, the British tend to use meetings to resolve issues, whereas, the French tend to use them to announce decisions. There is merit in the French approach. Have a private meeting (perhaps in a social environment) with your opposite project manager and manage the meeting to agree the decisions.

4.3   Presentations

4.3.1
See Section G, subsection 2, dealing with the format for a formal presentation to inform. A presentation to inform should be used for good news. You want everyone to receive the message without distortion. Consequently, it should be delivered to everyone at the same time.

4.4   Rumour

4.4.1
This is used to defuse or dissipate the effect of bad news. Let the bad news seep out in its worst form so that when the real situation is announced, people are relieved that it was not as bad as they thought.

4.4.2
Deliberate use of misinformation is even more dangerous. Nevertheless, it could be used in a tendering context by the responsible business development manager.

5   Social

5.1   Tea/Coffee Machine/Water Cooler

5.1.1
Make sure that you arrange to have tea and coffee facilities available in a convenient location. A lot of discussion, exchange of ideas, and status updating takes place in this informal environment. Drink more tea!

5.1.2
Visiting the social area should be part of your MBWA routine. See Section B Leadership and Motivation.

5.2   Canteen

5.2.1
You are fortunate if your company is far-sighted enough to have a canteen or other informal social relaxation areas. Most people will open up in this environment and talk about work during the lunch break. You will then have a management or problem-solving meeting every day!

6   Visual

6.1   Videotape

6.1.1
Used in the same manner as DVDs and CDs but is probably a redundant technology now – except in less-developed environments.

6.2   Photographs

6.2.1
Most events on a project are one-off occasions, and formal records should be kept for progress recording, potential publicity opportunities, and dispute resolution. Make sure that an appropriate spread of photographs is always taken from the same viewpoint(s) each month or week, for record purposes and for the construction manager's monthly report.

6.2.2
Informal photographs for newsletters and other purposes are taken as and when needed.

6.2.3
A display of photographs in the project conference room can help home-office personnel understand the needs of the site people.

7   Other Communication Tools

a. Standard documents
b. Filing system
c. Web pages
d. Public relations material
e. Open days

8   Translators

8.1

Over time, translators will start to influence an outcome because they become so involved that they believe they are a principal participant. One client insisted on sending two Japanese delegates who couldn't speak English to a training course. After the first half of the course, the simultaneous translator who accompanied them started to become a delegate and would answer questions in their own right before translating the question.

8.1.1
In a negotiation, they may start to cut their own deal if they have been hired locally. After all, they will be living, and possibly working, with your counterpart after you have gone. I had to call for a 'time out' without our translator on one such occasion.

8.2
If necessary, go over the detail more than once (see 10.2 below.) On one occasion a manufacturer had two items for sale that they were willing to sell for £1 million (times were hard). When they asked for the translator to clarify the customer's offer; the response was that, yes, the customer was happy to pay £1 million each!

8.3
The consequence is that you must have a member of your own team who speaks the language but keeps quiet about it (especially from the translator).

9   A Difficulty

9.1
Today's systems and communication processes mean that most, if not all, data and information are available to senior management before you have had time to analyse it.

9.1.1
The managing director of the U.K. office of one of the leading contractors told me that his biggest problem was that his bosses in the States had read all the data before he got to the office!

9.2
Consequently, if you are on a special assignment, agree in advance with senior management *what* and *when* you will provide the daily or weekly update on actions and decisions. Hopefully, this should stop them interfering in the day-to-day detail.

10   Some Reminders

10.1

In a legal dispute, all records can be obtained by the other party (including computer back-ups of deleted files) in a process called *discovery*. I always found the internal exchanges between groups most revealing.

10.2

If you don't understand something – ask. This is particularly important when someone expounds on a proposed strategy or a proposed deal at the end of a protracted negotiation. Ask the person to explain it again, and if you still don't understand it clearly and you have confidence in your own intelligence – reject it.

Section B   Leadership and Motivation

"Leadership is that combination of persuasion, compulsion and example that makes men do what you want them to do."
Field Marshall Slim, 1962.

"Never tell people how to do things. Tell them what to do and they will surprise you with their ingenuity."
George S. Patton.

"Being powerful is like being a lady. If you have to tell people you are, you aren't."
Margaret Thatcher.

Remember that you are managing in a matrix organization. You, the project manager, define *what* you want done, and the functional managers are responsible for *how* the technology is performed. Thus, you have to lead and manage people over whom you have no authority.

> The management team or contractor's preferred project management structure will tend towards a strong matrix/task force. This option is probably necessary in order to overcome the disadvantages of the design office environment. In the design office no one works directly for the project manager and they are surrounded by equally (or more) senior functional managers. Their power and influence is consequently somewhat diminished.
> Where the pure task force does exist is at the installation group level. The installation manager does have responsibility for hiring and firing the resources and consequently they are seen to work for the manager more directly. This greater power base may explain why the construction management organization is often seen as more effective than the project management organization.[3]

The traits theory of leadership has not stood the test of time. Research has demonstrated that leadership is about behaviours and processes people use to direct, control, guide, and inspire others. Leaders, then, are not born but made. Because leadership is an influencing process between people, the behavioural approach to leadership assumes that subordinates will work better for managers who use certain styles of leadership than those who employ different styles.

   This section summarizes a number of leadership and motivation models that are applicable in a project context. The contention is not that they are competing models but that each has a role to play at a particular stage of project development or in a particular context.

---

3 'Project Organization Structures from Logic to Reality' by Garth G. F. Ward. Paper presented to the Norwegian Institute of Technology Nordnet '91 conference, Trondheim, Norway, 3–5 June 1991.

1 Consensus to Dictatorial Continuum by Tannenbaum and Schmidt

This is a practical one- dimensional continuum model. It illustrates the relationship between the levels of freedom that a manager chooses to give to a team and the level of authority used by the manager. The styles of management behaviour range from subordinate-centred (relationship orientated) to manager-centred (task orientated), see Figure VI.B.1.

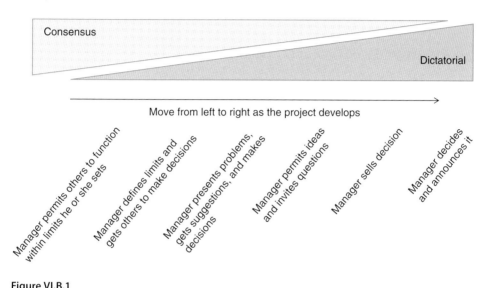

Figure VI.B.1

1.1

This is a useful model to have in mind at the start of the project when thinking of the whole project group. It is important to get the team to buy in and be committed to the approach to be used on the project. Thus for project launch, we need more freedom for the project team to achieve a consensus. As the work progresses, the project manager can be more directional. Having bought in, the team wants decisions and will be more willing to accept a directional leadership style, with the project manager exerting extra authority.

1.2

Do not try to get everyone to reach agreement. Seeking a total consensus can lead to endless haggling and conflict. Margaret Thatcher felt that "consensus doesn't give any direction."[4] People will accept a decision that they do not fully agree with if the process (see a – d below) by which decisions are reached is seen to be fair. By having their say and their opinions considered seriously, they will accept a 'consensus

---

4 From a speech in the House of Commons on Wednesday, April 10, 2013.

with qualification'[5] result much more quickly. On the other hand, all Churchill wanted was acceptance of his views after reasonable discussion.

a. Each person must be free to suggest alternatives.
b. The group fully discusses the arguments for and against each option.
c. The group collectively modifies an alternative, as needed, to improve it.
d. The group or manager selects the option most acceptable to the group. The option is the one that enables individuals to say: "I can live with it."

See also Section K, subsection 3, Binary Decision-making.

2 The Three S's of Group Communications

This model[6] provides a useful guide on how to apply the Tannenbaum and Schmidt's continuum to a group of people.

2.1

With a small group: the interpersonal skills of the group as a whole will be relatively high. Thus, a decision-sharing style is workable (shown by the full line from bottom left to top right in Figure VI.B.2).

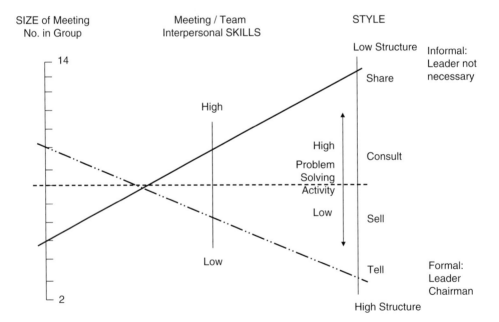

Figure VI.B.2

---

5 Kathleen M. Eisenhardt, Jean L. Kahwajy, and L.J. Bourgeois III, 'How Management Teams Can Have a Good Fight,' *Harvard Business Review* July–August 1997.
6 My friend and colleague Roger Griffin, sometime training manager of Bechtel Ltd., gave me a sketch of this diagram but was unable to trace its origins. I have added the three S's.

2.2

With a bigger group: the interpersonal skills of the group will be reduced. Consequently, a more formal, but still consultative style will be required (shown by the horizontal dashed line in Figure VI.B.2).

  With a large group: the interpersonal skills of the whole group will be relatively poor. Hence a more structured approach directing the group will be needed (shown by the chain link line from top left to bottom right in Figure VI.B.2).

3   Situational Leadership by Kenneth Blanchard and Dr. Paul Hersey

Situational leadership theory states that leaders should change their leadership styles based on the situation. This model folds the Tannenbaum and Schmidt straight-line continuum through 90 degrees to create two dimensions. The Y axis indicates how much relationship and supportive behaviour is required, and the X axis indicates the directive effort needed to achieve the task. The insightful part of the model is that it says that how much of the one or other dimension the leader needs to apply depends

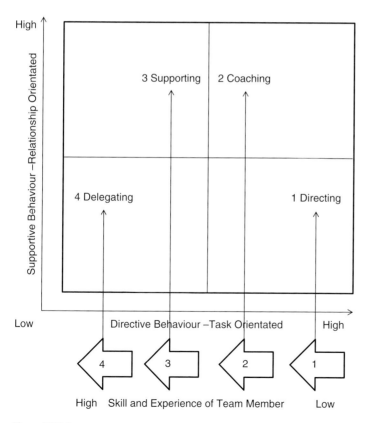

Figure VI.B.3

on the development level of the follower. The needs of the situation depend upon the commitment, motivation, or maturity of the person being led and their knowledge, experience, or competency/capability for the task. Using this theory, leaders should be able to place more or less emphasis on the task and more or less emphasis on the relationship with the person they're leading. A particular leadership style may be used on one task and a different style used with the same person, on a different task.

3.1 Supervising Styles

a. 1 Directing: The supervisor tells the team member exactly what to do and how to do it.
b. 2 Coaching: The supervisor still provides information and direction, but there is more communication with the team member. The supervisor 'sells' their message to get the person on board.
c. 3 Supporting: The supervisor focuses more on the relationship and less on direction. The supervisor works with the person and helps with decision-making responsibilities.
d. 4 Delegating: The supervisor passes most of the responsibility onto the team member. The supervisor still monitors progress at key points, but they are less involved in decisions.

Don't forget that when you delegate, you are still accountable. See subsection 7 below.

3.2 Skill/Experience and Motivation

a. 1 Low skill, high motivation – Generally lacking the specific skills required for the job in hand but has the confidence and/or motivation to tackle the task. It's a new experience, and the person is keen to get on with it.
b. 2 Some skill, low motivation – May have some of the skills needed but won't be able to do the job without help. The task or the situation may be new to them, but the work is not as interesting as they thought.
c. 3 High skill, variable motivation – Experienced and capable but may lack the confidence to go it alone or the motivation to do it well/quickly. They need encouragement to keep at it.
d. 4 High skill, high motivation – Experienced at the job and comfortable with their own ability to do it well. May even be more skilled than the supervisor. You can leave them to get on with the work.

3.3 Time Management

The heavy emphasis on relationships, coaching, and support means that they are very time-consuming for the supervisor involved. Consequently, once the team has bought in, more use should be made of directing and delegating since this frees up the manager's time.

4 Task, Team, Individual – Action Centred Leadership by John Adair

John Adair's Action-Centred Leadership model for team leadership and management (see Part IV, Section A Project Launch, Figure IV.A.2) provides a focus for the three core management responsibilities:

- Achieving the task
- Managing the team or group
- Managing individuals

Identifying and grouping some key concepts can also make it an effective model for motivation.

4.1 Task focus

a. For the project team to be motivated, it is essential for the scope to be comprehensively and fully defined. In addition, the individuals need to understand the significance of their task and how it fits into the overall project plan.
b. People like to achieve targets. Consequently, milestones can be used as motivators for the team. Providing 'stretch targets' (the early finish date from the network analysis) for specific tasks similarly challenges individuals.
c. As the project manager, you will be arriving early and leaving late! Be punctual for meetings, and start on time in order to develop a productive project culture.

4.2 Team focus

a. The capability of the team will depend on the relationships developed during team building (see Part V, Section Q, subsection 2.) and the clarity of the roles and responsibilities. Having said this, in a truly effective team, the individual team members help each other regardless of their job descriptions. Further, when and where appropriate, the project manager should allow the leadership of the team to devolve to other team members.
b. It is useful if a bonus scheme has been negotiated to achieve project cost, time, and quality targets. This generates a competitive element for the team.
c. Try and eliminate your 'project management ego.' Talk about the team as a team and keep everyone updated on developments in the project.
d. The power of a team still continues to amaze me. Use the creative and constructive ability of the group process.

4.3 Individual focus

a. The individual team members should be selected on the basis of their being able to satisfy their personal agenda. This will depend on the stage of development of their careers and may be difficult to identify.
b. Job titles are cheap. Where possible, use them to satisfy individual career development ambitions.

   c. Experienced individuals will not like being told what to do. Give them tasks that utilize their experience and skills. Consequently, develop individuals by making more use of the delegating style (see paragraph 3.1d above). Senior management recognition of your development of individuals will be to your own benefit.

   d. If the bonus scheme, mentioned above, has an element that is distributed to individuals, it is even more powerful as a motivator.

   e. As indicated in the situational leadership model, it is sometimes necessary to support and encourage individuals.

   f. I once bent the rules to benefit someone's personal needs. From that point on, I was taken advantage of. I soon changed to my construction manager's tough line: Be *firm, fair,* and *consistent,* and people will respond positively.

   g. A colleague once said: "That was well done, Garth." Even though, I was 100 per cent certain that I was being manipulated in the particular circumstances, I felt good about it. It made me realise how powerful acknowledgement of good work can be.

5 Leadership and Management Roles

5.1

Adizes[7] recognizes that the management process is too complicated for any one individual as effective management entails simultaneously performing four individually conflicting management roles:

   a. *The Producer*-is oriented towards creating results and has a thorough knowledge of their field.

   b. *The Administrator*-is oriented towards planning and scheduling.

   c. *The Entrepreneur*-is oriented towards generating new ideas and plans of action.

   d. *The Integrator*-is oriented towards turning individual goals into group goals.

5.2

Adizes argues that, in practice, no single individual is capable of performing all four of the necessary managerial roles at the same time. Adizes's first two roles may be seen as primarily associated with the concept of management (efficiency). Whereas, roles 3 and 4 are more closely connected to the concept of leadership (effectiveness).

5.3

There are interesting correlations with different aspects of Belbin's team roles (see Part V, Section R). For example: 'the producer' is Belbin's company worker/implementer and 'the integrator' is Belbin's team worker.

---

7 Dr. Ichak Adizes. *How to Solve the Mismanagement Crisis.* Published by the Adizes Institute. 1979.

6 Management by Walking/Wandering Around MBWA[8]

6.1

MBWA is an informal and unstructured approach to involvement by the manager in the work of subordinates. In practice the project manager makes informal visits to designers and draughtsmen (and foremen on site) and listens to their comments and complaints about their work. The process enables issues and suggestions to be raised that might not otherwise get an airing.

6.1.1
MBWA is necessary because you cannot understand what goes on in the mind of some-one more than two hieratical levels below you (or above you).

6.1.2
On one occasion I found that a draughtsman was working on a different project. It was not that they were moonlighting, but that the message from the top had been badly distorted by the time it had been received. In this situation I made no comment but told the supervisor that I thought that there was a problem and advised them to look into matters. I believe that I acquired the nickname Snoopy. The process works.

6.1.3
Some of the more insightful observations of what is occurring in an organization are obtained by exchanges with acquaintances on the way to or in the loo!

7 Responsibility

7.1

This is what the father of the U.S. Navy nuclear propulsion programme, Admiral Hyman G. Rickover, had to say on the subject of responsibility[9]:

7.1.1
"Responsibility is a unique concept: it can only reside and inhere within a single individual. You may share it with others, but your portion is not diminished. You may delegate it, but it is still with you. You may disclaim it, but you cannot divest yourself of it. Even if you do not recognise it or admit its presence, you cannot escape it. If the responsibility is rightfully yours, no evasion, or ignorance or passing the blame can shift the burden to someone else. Unless you can point your finger at the man who is responsible when something goes wrong, then you have never had anyone really responsible."

---

8 Ed Carlson of United Airlines used *Management by Walking About,* and William Hewlett and David Packard, founders of Hewlett Packard (HP), used *Management by Wandering Around* in their companies. Tom Peters and Robert Waterman Jr in their book *In Search of Excellence,* 1982, included lessons learned from HP and other companies that used a similar style – and the term MBWA was born.
9 This quotation was sent to me by Charlie R. H. Field, Lieutenant Commander RN, who attended a Cranfield project management course.

8   Leadership – More Than a Management Model

It has been stated that leaders are made rather than born. The various leadership models demonstrate that leadership can be learnt. But, there must be more to it than just a management model? See the edited extracts from the article below.

8.1

Know Your Men, Your Business and Yourself.[10]

You will have in your charge loyal but untrained citizens, who look to you for instruction and guidance. Your word will be their law. Your most casual remark will be remembered. Your mannerism will be aped. Your clothing, your carriage, your vocabulary, your manner of command will be imitated.

When you join your organization you will find there a willing body of men who ask from you nothing more than the qualities that will command their respect, their loyalty and their obedience. They are perfectly ready and eager to follow you so long as you can convince them that you have those qualities.

Leadership is a composite of a number of qualities. Among the most important I would list self-confidence, moral ascendancy, self-sacrifice, paternalism, fairness, initiative, decision, dignity, and courage.

Self-confidence results, first, from exact knowledge, second, the ability to impart that knowledge. … Men will not have confidence in an officer unless he knows his business, and he must know it from the ground up.

And not only should each officer know thoroughly the duties of his own grade, but he should study those of two grades next above him.

Self-sacrifice is essential to leadership. You will give, give all the time. You will give yourself physically, for the longest hours, the hardest work and the greatest responsibility is the lot of the captain. He is the first man up in the morning and the last man in at night. He works while others sleep.

Fairness is another element without which leadership can neither be built up nor maintained. There must be that fairness which treats all men justly. I do not say alike, for you cannot treat all men alike – that would be assuming that all men are cut from the same piece.

When one of your men accomplishes an especially creditable piece of work, see that he gets the proper reward. Turn heaven and earth upside down to get it for him. Don't try and take it away from him and hog it for yourself. You may do this and get away with it, but have lost the respect and loyalty of your men. Sooner or later your brother officers will hear of it and shun you like a leper. … Give the man under you his due.

When an emergency arises, certain men calmly give instant orders which later, on analysis, prove to be …very nearly the right thing to have done. …You may say "That man is a genius. He hasn't had time to reason … he acted intuitively." Forget it "Genius is merely the capacity for taking infinite pains." The man who was ready is the man who has prepared himself. He has studied beforehand the possible situation that might arise, he has made tentative plans covering such situations.

---

10 Extracts from a speech given by Major C. A. Bach, U.S. Army in 1917 to the graduating officers of the Second Training Camp at Fort Sheridan. The Waco (Texas) *Daily Times Herald*, learning of the great interest the speech had aroused, obtained a copy and printed it verbatim in January 1918. A copy of the speech was inserted in the Congressional Record by Senator Henrik Shipstead of Minnesota in November 1942 and printed as Congressional Document 289.

Any reasonable order in an emergency is better than no order. The situation is there. Meet it. It is better to do something and do the wrong thing than to hesitate, hunt around for the right thing to do and wind up by doing nothing at all. And, having decided on a line of action, stick to it. Don't vacillate. Men have no confidence in an officer who doesn't know his own mind.

Moral courage demands that you assume responsibility for your own acts. If your subordinates have loyally carried out your orders and the movement you directed is a failure, the failure is yours, not theirs.

Furthermore, you will need moral courage to determine the fate of those under you. … Keep clearly in mind your personal integrity. … if you are called for a recommendation concerning a man whom, for personal reasons you thoroughly dislike, do not fail to do him full justice. Remember that your aim is the general good.

And lastly, if you aspire to leadership, I would urge you to study men. Get under their skins and find out what is inside. Some men are quite different from what they appear to be on the surface. Determine the workings of their minds."

8.2

Know your team, know your project, and know yourself.

9   Thoughts for the Day[11]

   a. The boss says, 'Go'; the leader says 'Let's go' Harry Gordon Selfridge (HGS).
   b. The boss depends on authority; the leader depends on goodwill (HGS).
   c. The boss drives people; the leader coaches them (HGS).
   d. The boss inspires fear; the leader inspires enthusiasm (HGS).
   e. The boss says 'I'; the leader says 'We' (HGS).
   f. Manage the present, but plan for the future.
   g. Be ambitious about beating targets.
   h. Achievement is born from challenges.
   i. Only your effort can develop *your* potential.
   j. Success and failure can be equally instructive.
   k. Success comes to those who *want* it – seek it out.
   l. You can learn from failure but only if you own up to it.
   m. If you don't make sure it's going right, you can be sure it will go wrong.
   n. Just because things are not going the way you planned doesn't mean that what you are doing is wrong and that it won't be a success.
   o. Just keep going, and you will reach your goal.
   p. Great results are the result of attention to detail.
   q. Your attitude determines your success.
   r. Keep an open mind, and keep asking why.
   s. Maintaining the status quo is the beginning of decline.
   t. Give back more than you receive.
   u. Be more concerned by the lack of ideas rather than lack of success.

---

11 These have primarily been taken from *The Matsushita Perspective – A Business Philosophy*. PHP Institute, Inc. 1997. Some modifications have been made, and one or two additions have been included. Other quotations have been identified.

    v. Without obstacles to overcome, there can be no satisfaction.

    w. A complaint is an opportunity to build relationships.

    x. Above all, be a good listener.

    y. Don't panic, no matter what.

    z. Trust breeds trust.

  aa. Strength comes from the wisdom of many.

  bb. Just as a project needs procedures, an individual needs principles.

  cc. Play to your strengths if you want to succeed.

  dd. Luck is the by-product of effort.

  ee. Cheer the success of others, but do your own thing.

   ff. Everyone has talent – it only needs to be uncovered.

  gg. Focus on what people can do, not what they can't.

  hh. Be prepared for the unexpected.

   ii. Continuous daily effort makes the difference.

   jj. Information comes to those who seek it.

  kk. Don't look for reasons why something can't be done; find a way it can.

   ll. Do what you know needs to be done.

 mm. If something is really important you should spend at least 50 per cent of your time on it (Tom Peters).

  nn. If you don't understand something, ask for it to be explained.

  oo. Ask for advice, but do what you think is right.

  pp. Your experience is your most valuable asset.

  qq. It is dangerous to be right in matters on which the established authorities are wrong (Voltaire).

   rr. Never confuse motion with action (Benjamin Franklin).

  ss. Until one is committed, there is hesitancy, the chance to draw back, always ineffectiveness (Goethe).

   tt. Whatever you can do or dream, begin it! Boldness has genius, power, and magic in it. Begin it now! (Goethe).

  uu. Perfection doesn't exist but the evolution towards it does. (Enzo Ferrari).

Section C   Managing and Conducting Meetings

"The mutual confidence on which all else depends can be maintained only by an open mind and a brave reliance upon free discussion."
Speech to the Board of Regents, University of the State of New York. Learned Hand, 1950s.

After the Duke of Wellington had held his first cabinet meeting as prime minister, he is said to have exclaimed: "An extraordinary affair. I gave them their orders and they wanted to stay and discuss them."

1   Planning the Meeting

1.1

Firstly, is a meeting the best way to accomplish the defined purpose? If yes, then what process will best serve the purpose?

1.1.1

Spend some time deciding why the meeting is necessary. In a project most meetings will have been predetermined, for example, progress meetings or the project manager's meeting. These meetings will have been entered on the project calendar for a particular time and day, either weekly or monthly. So there is no excuse for non-attendance. For extraordinary meetings, say, problem-solving meetings, make sure that people have sufficient advance notice.

1.2

Be clear about your objectives and the results you want to achieve from the meeting. Make sure that everyone understands the real goal of the meeting. For example, the real goal may not be to solve the problem but to get the authorised manager to agree with the team's conclusion such that they will sign off on the proposed course of action.

1.3

Issue an agenda to the participants. Can any agenda items be eliminated by brief one-to-one conversations or a chat on the telephone?

1.4

Send a memo with the agenda telling people when and where the meeting is to take place and tell them what data, documents, and information may be required. Inform people in sufficient time before the meeting to enable people to prepare for their contribution.

1.5

Attendance at meetings should be at the lowest supervisory level empowered to make decisions. Thus, only problems that cannot be resolved at this level are passed up the next supervisory level. Similarly, only problems unresolved at this intermediate level are passed up to the project manager's meeting level.

1.6

Remember, when people are at a meeting, they are not working. Do not let people attend who are not really concerned with the objective of the meeting. They may just be looking for a reason to book man hours on their timesheet.

1.7

Try not to cancel a meeting; the lost man hours and disruption involved in rescheduling are not worth the trouble. If you cannot attend, deputize a senior member of your team to take the meeting on your behalf. Similarly, if someone else cannot attend, ask them to nominate a deputy.

2   The Agenda

2.1

Develop the agenda, listing the subjects in a logical order, grouping similar and related subjects. The agenda should not be just a list of topics, but each item should explain the circumstances of the issue and what needs to be accomplished. Develop a standard agenda for progress meetings.

2.2

Allocate some approximate times for discussion. Place a time limit on urgent items of low importance. This will enable the important items to have the maximum amount of time for discussion.

2.3

Fix the duration of the meeting, appropriate for the agenda, and announce the finishing time (this can be offset by the discussion expanding to fill the time available). Nevertheless, people are more concise and to the point if there is a demanding deadline.

2.4

Start meetings one hour (or whatever duration you have fixed for the meeting) before lunch or one hour before the time to go home. That way people are really serious if they want to expand on a topic.

3 Manage the Process and the People

3.1

Try not to have too many participants. In a project this can sometimes be difficult to achieve. Consequently, consider asking people to attend the progress meeting in stages, rather than sitting through long discussions that do not concern them.

3.2

The seating layout is important; arrange the seating so that everyone gets to see every-one else. As the chairperson, sit at the head of the table. If you know that there is an aggressive person, ask them to sit next to you (or in other circumstances go and sit next to them), not opposite them. Sit aggressive people on the same side. It is much more dif-ficult to have an argument with someone sitting next to you rather than opposite you. I've tried it, and it works.

3.3

Introduce people (names and project roles) so that everyone knows who all the atten-dees are.

3.4

Start as you mean to go on. Do not wait for latecomers. Be prepared to take dramatic action for persistent latecomers or for people who come unprepared. Take a clock to the meeting, and charge their department for any time wasted!

3.5

The chairman sets the tone and process of a meeting. In a progress meeting, they will direct the use of problem-solving and decision-making techniques as needed (at the autocratic end of Tannenbaum and Schmidt's continuum; see the previous Section). Whereas for a problem-solving meeting, they will adopt a neutral position and remain impartial (at, or towards, the consensus end of Tannenbaum and Schmidt's continuum). They will defer any judgment and avoid appearing to take sides. They will protect airtime, the acceptability of ideas, and the consideration of alternatives from loss or rejection without full discussion of the group. They will actively encourage involvement but regulate input from those who would monopolize the discussion.

3.6

The positive personality can be of great help and should be used, but don't let a forceful character monopolize the discussion. Stop dominant personalities taking control. Use questions to defuse any aggression and remind people what the objectives are. Remind them that we are all part of the same project team. However, aggressive people will want to have their say on a particular matter, so let them. Then ask the quiet ones what their views are.

3.6.1

An opposite approach is to ask the quieter members to give their opinions first, when a 'senior person' or 'expert' might repress the other members at the meeting and so stifle their contributions. Keep an eye on the hesitant, quieter people. When there is a suitable opportunity, ask one of them by name an easy question that they can answer well. Thank them for their opinion.

3.6.2

Acknowledge the experience of the person who is being uncooperative and let them feel that you depend on their help for the success of the meeting.

3.6.3

The talkative person may be of real value if they are well informed and so should not necessarily be discouraged. However, if they are allowed to go on for too long, they may get boring. So, interrupt politely and ask a direct question of someone else.

3.6.4

The 'academic' type needs to be kept to the point, and it may be necessary to paraphrase what they have said for the benefit of the rest of the meeting.

3.6.5

The person who knows it all may not know it all. Ask them to back up their opinion with reasons. If they are valid, move on. If they are faulty, ask others to comment.

3.6.6

The person who is constantly asking questions may be trying to catch you out. Pass the questions back to the meeting, and then ask for their own views.

3.6.7

If someone is disinterested, occasionally ask for their advice based on their experience. Try and give the impression that their opinion is needed.

3.6.8

Don't allow anyone to become personal.

3.7

In a progress meeting, the participants should know the type of information needed. However, they will need to respond to questions concisely and correctly. Try to avoid suppressing contributions.

3.8

In a problem-solving meeting, people need to participate, get involved, and share information and ideas freely. They must defer all judgment until ideas and suggestions have been fully considered.

3.9

A requisite skill required of people is listening. If everyone is to have a say, then people will be required to listen for most of the time.

3.10

Be observant and check that people have understood the issues and conclusions.

3.11

See Section E, Personal Skills, subsections 1 to 8 Interactions with Others.

4 Control the Discussion

4.1

Remind everyone what the meeting relates to and what the objectives are (see 1.2 above).

4.1.1

Explain why each topic is being discussed as the meeting progresses, and make clear the result that you hope to achieve from each part of the discussion.

4.2

Margaret Thatcher started every meeting with: "What are the facts?" Facts trump opinions.

4.3

As in a project, structure the discussion in phases: collect all the information and facts and then analyse the data and information. Then, using the synergy of the group process, construct decisions about the action to be taken.

4.4

Stop people changing the subject or going back and revisiting something that has already been decided.

4.5

Another danger is people raising irrelevant issues and wandering onto other topics on the agenda. Do not allow excessive discussion on unimportant detail. Bring the meeting back to the issue being dealt with.

4.6

Stop people having separate discussions/meetings. Ask the people concerned to share the issues with the whole meeting.

4.7

Clarify obscure statements. If you don't understand what someone has said, ask them to explain it again in a different way.

4.8

When generalisations are used; ask the meeting if it is everyone's experience.

4.9

Remain objective and listen to both sides of an argument. Consider throwing questions back to the meeting before expressing your point of view. Don't prejudice the discussion by imposing your point of view. Keep your views to a minimum.

4.10

Try to make sure that every contribution is relevant and moves the discussion towards a resolution.

4.11

As the meeting progresses, it is useful to check with the group on how well the meeting is achieving its goals.

4.12

Tell the meeting when it is approaching the end, and decide what will be the best use of the remaining time.

4.13

Do not let your chairmanship become oppressive by controlling too tightly.

5 Construct Decisions and Summarize

5.1

Let everyone have their say. Pull together similar contributions. Check around the meeting to make sure no one has been left out.

5.2

When brainstorming (see Section J, subsection 1), record all ideas and suggestions.

5.3

Generate ideas from each other. Look for alternatives – don't undermine people or their ideas.

5.4

Combine ideas into an acceptable solution. Don't chase a consensus; it is too time consuming (see Section B, Paragraph 1.2). Do not let the pressure to get a unanimous agreement create a groupthink decision.

5.5

Summarize and express things in a different way, but don't elaborate excessively. Check that everyone still understands the issue.

6   Record and Notify

6.1

Organize someone (from the project controls group) to take notes during the meeting, preferably someone who can write up the minutes, briefly and unambiguously, as you progress.

6.2

The minutes should be in a consistent format that should be the same for all meetings. They should be complete in themselves and preferably not refer to other documents.

6.2.1

Record all decisions in the minutes and the names of the people responsible for the actions. Set target dates for when action items are to be completed.

6.3

Make sure that you review the minutes and edit them as necessary. The person taking the minutes is responsible for ensuring that the minutes are agreed by the parties involved and issued within a day or so.

6.4

Distribute copies of the minutes to the attendees and in accordance with the distribution matrix. Highlight to individuals the actions that they are responsible for.

6.5

Remember that minutes of a meeting with a client are agreements that are contractually binding unless you have made an agreement to the contrary. In which case, the published minutes will need a statement on each page stating that they 'do not intend to form a contract.'

6.6

After difficult meetings, you may overhear negative mutterings as the attendees leave the room. Alternatively, dissenters during the meeting may later renege on the agreements made. In these circumstances, you may need to get everyone to *sign off* on the agreements made.

Section D   Negotiation

"Trades would not take place unless it were advantageous to the parties concerned, of course it is better to strike as good a bargain as one's bargaining position permits. The worst outcome is when, by over reaching greed, no bargain is struck, and a trade that could have been advantageous to both parties does not come off at all."

Benjamin Franklin.

This section is really only 'a starter for ten.' Read the books referenced in the section.

We all negotiate, and all negotiators use the same skills. Negotiation is about resolving differences between people whether it is to get a salary increase, obtain additional resources, buy materials, or finalize a contract for a project. Nevertheless, many or most people are afraid to negotiate, or it is discouraged by company procedures. However, we don't practice it enough. In any situation, the parties have a different point of view, and there is a natural tendency to take a particular position and argue for it, make concessions, and settle for a compromise. However, arguments cannot be negotiated; only proposals can. Giving in, in stages, isn't negotiating.

All good negotiations have a structure and the following eight-step approach to negotiating has been extracted from the Rank training video 'The Art of Negotiating'. A few additions from the book *Getting to Yes*[12] have been included. The video was based on the book *Managing Negotiations – How to Get a Better Deal*[13] This book was the first to put negotiating into a structured process, and in the video, this is condensed into four phases as follows:

Phase 1	Prepare
Phase 2	Discuss Signal
Phase 3	Propose Package
Phase 4	Bargain Close Agree

1   Preparation for Negotiation

a. Objectives:
  i. Have different priorities. Identify your *LIM*: *like* to have, *intend* to have and *must* have.
  ii. Identify what is to be achieved.
  iii. Any additional items or services required or to be provided.
  iv. Any factors dependant on the circumstances.
b. Information:
  i. Identify what is negotiable – in fact anything and everything.
  ii. Gather the facts.

---

12 *Getting to Yes – Negotiating an agreement without Giving In* by Roger Fisher and William Ury, 1982. ISBN 0-09-92484-5; 2nd ed. by Roger Fisher, William Ury, and Bruce Patton. Arrow Books Ltd, 1997. For people wishing to take the subject further read: *Getting Past No – Negotiating Your Way From Confrontation to Cooperation*, by William Ury. Bantam Books, 1991. ISBN 0-553-37131-2.
13 *Managing Negotiations – How to Get a Better Deal* by Gavin Kennedy, John Benson and John McMillan. Hutchinson Business Books, 1980. ISBN 0-09-168891-4.

      iii. Identify strong points and weaknesses.

      iv. Get to know about the other party – the company and people.

      v. Separate the people from the problem.

c. Team Roles:

      i. Decide on the team leader and team roles and allocate tasks.

      ii. The lead negotiator, who indicates when others may speak, may not be the team leader but will summarize what they have understood.

      iii. The listener and analyser of what the other party are proposing.

      iv. The observer of the other people's reactions, body language, and eye blink rate.

      v. A note taker.

      vi. A specialist may be required in technical negotiations. They need careful briefing; they have a tendency to give things away!

d. Strategy

      i. Should be simple and flexible.

      ii. Assess how the other party will view the problem.

      iii. Prepare a list of points to be discussed in a logical sequence.

      iv. Provide an agenda to the other party.

      v. Identify concessions you might make and what you will require in return.

      vi. Aim for a 'win-win' situation.

      vii. Identify your fall-back position. Know your BATNA – your *best alternative to a negotiated agreement*.

      viii. Resolve technical issues first.

      ix. Consider 'home', 'away', or 'neutral' location.

      x. Be clear about your limits of authority. Remember you do have authority to make recommendations.

      xi. Are there any security considerations?

      xii. Rehearse!

2 Discuss Interests

a. Listen don't assume – test assumptions.

b. Exchange information – ask questions.

c. Focus on and explore interests – identify resistance and inhibitions.

3 Signal

a. Open realistically.

b. Listen and watch for signals. Have they signalled? Have you?

c. Recognise and confirm messages.

d. Develop ideas.

e. Avoid irritations and displays of emotion.

3.1

It can be useful to have a break, a 'time out', in preparation for 4.

4   Propose for Movement

a. Negotiating is about mutual movement Proposals advance negotiations.
b. Use 'What if', 'could consider', and 'maybe'.
c. Try to satisfy the other party's needs.
d. What is it worth to the other person? Beware of assumptions.
e. Try and enlarge the 'pie'. Invent options for mutual gain.
f. Listen, don't interrupt, question, clarify, summarize, and respond with a counter proposal.

4.1

An alternative 'time out' in preparation for step 5.

5   Package

a. Express the proposal in a different form, which addresses the interests and inhibitions of the other party.
b. Insist on objective criteria that are independent and practical.
c. Use If … Then … statements. If you hear these words, you may be dealing with trained negotiators.

6   Bargain

a. Identify conditions before making offers.
b. Do not give concessions without getting something in return. Always get something in exchange.
c. Give the other party what it wants/values, if it is cheap to you; in exchange for something valuable to you.
d. Aim for a '*win-win*' situation.
e. Silence and time-outs can be effective ploys.

7   Close the Deal

a. Close realistically,
b. 'If you do that … then …. we have a deal.'
c. The final concession traded for an agreement.
d. If you propose doing business again with the same people or organization, it is important to consider how the final agreement will affect relationships.

8   Agree the Deal

a. Agree what has been agreed. Both parties should feel happy with the agreement.
b. Summarize and record the agreement.
c. Agree on an action plan and implementation process.

9   Techniques and Tricks

9.1

Even if you would not dream of using tricks, it is important that you recognize them when used by others.

   a. Use probing questions to discover the other party's weak points. Keep the questions short: how and why? Would you explain your cost build up?

   b. Ask/find out where they have travelled from. This gives some indication of how much they want the order.

   c. Impose time pressure by offering (right at the start) to confirm their return air-flights.

   d. Give the impression that you have a strong position by indicating that the negotiation must be completed by a certain time.

   e. Use pressure: advantages of rank, name dropping, or overt signals of exasperation.

   f. Say as little as possible forcing the other party to do the talking and refuse to be drawn.

   g. Make out that you didn't understand a point, thus forcing the other party to clarify their point and perhaps provide additional information.

   h. Keep people waiting. Delay progress until the timing is more favourable.

   i. Flattery can be effective. Ask for their opinion or advice about something.

   j. Deliberately over or understate some point in order to provoke a disclosure of the other side's point of view.

   k. Make a credible low offer and imply that you don't see the need to move from the position.

   l. Act as an agent working within limits and then say, "I'm sorry, but that is as far as I can go."

   m. Hard/soft approach – one person is deliberately difficult and when they leave the room, the second team member emphasizes how reasonable they are.

   n. Pre-arrange an interruption in case matters go badly. You can then withdraw and work out a new approach.

9.2

If you are under pressure, absorb it by taking notes.

9.3

Keep all unsettled issues linked, and keep them until the end.

9.4

Nothing is agreed until everything is agreed.

9.5

The nibble: having virtually agreed the deal, you suddenly say: "Oh, I forgot. Can we …?" You then increase the price slightly or ask for something extra. It may not win you friends, but it works.

9.6

Time out: During the time out, leave one of the team members behind in the room to chat (socially) to the other team whilst you rethink your strategy.

Section E Personal Skills

"Non omnia possumus omnes." – We are not all capable of everything.
Eclogues VIII, Virgil, 70 – 19 B.C.
So, make the most of what you have, and improve what you can.
Garth Ward, 2014.

The focus of this section is on interactions with others, since this is the only mechanism the project manager has to get things done.

> The most important skill in project management is to communicate. When a team can do that properly it has the possibility of being a great success. Communication is a two way thing. There is the giving it out bit, but there is also the getting it in bit, that is, the listening. One only really learns how to communicate, when one has learnt how to listen and, particularly, how to hear what is not being said.[14]

Relationships are the most important factor in communications. Without good relationships, people will be reluctant to talk to you. A conversation is all the project manager has in their tool bag. Consequently, the quality of the conversation has a direct impact on the result.

I have never forgotten visiting a project manager who was just finishing a conversation with a member of his project team. The closing remark was: "Discuss it with Fred." To which the person replied, "Who's Fred?" The PM responded, "He's the person who sits next to you!"

Eliminate the office door as a barrier between you and the team. Maintain an 'open-door policy'.

See Section A for Communication Mechanisms and Section C Managing and Conducting Meetings.

1 Planning an Interaction with Others

1.1

Pick the time and location for your interaction with other people and make sure that you are clear about its purpose.

1.2

Treat important conversations with the client as you would a project. Think about what you hope to achieve and then plan it. You may wish to even go through it with your project controls manager and get them to be a 'devil's advocate.'

---

14 Pat McHugh, 2012.Written specifically for this book.

1.3

If you need to talk to someone in your office on a one-to-one basis, shut the door. Don't let interruptions or telephone calls take precedence (as in India or Egypt) and, as I discovered on my return on one occasion, so do the British.

1.4

Delegate telephone calls to your secretary/admin assistant or the deputy project manager. If you do not have any of these people, delegate to one of the project engineers.

1.4.1

Otherwise, if you can, tell people that you will ring them back. Make a list of all your calls and call people back half an hour before lunch or leaving time!

1.5

Think of the individual(s) and think about expectations. Ask yourself:

a. 'What do I expect to deliver?'
b. 'What do I expect to get back?'
c. 'What does the other person expect of me?'
d. Identify priorities and the information required.

2   The Exchange

2.1

Tell it as it is. Don't be too British and say something is quite good; if it is good, just say it is good. If it's not really what you wanted, say it's not. Be more like the Dutch, who can be perceived by the British as being rude when in fact they are being open and honest.

2.2

Try not to interrupt people. Control your desire to jump into a conversation with your point of view. Listen to what they have to say and pause before giving a reactive answer. If necessary say, "I need to think about that. I will get back to you."

2.3

If you want to command attention in a meeting, and make people pay attention, lower your voice rather than raising it (in competition with others).

2.4

You don't have to win every argument – maybe the other person has a valid point to make.

2.5

See Section C Managing and Conduction Meetings, paragraph 3.6.

3   Asking Questions

3.1

Avoid questions that invite a yes or no answer. Ask open questions that encourage people to contribute their own ideas or point of view. "What makes you think that you are a good project manager?" "Tell me about …"

3.2

Closed questions are used to enquire about specific facts. "How many projects have you completed on time and within budget?" "What is the status of the activity?"

3.3

Check or clarify information with probing questions. "How did you manage to …?" "Tell me what happened next …?"

3.4

Ask leading questions to get acceptance of your point of view. "You agree then, that time is more important than cost?" "You don't disagree with what has been said, then …?"

3.5

Pose hypothetical questions about "If you were the client, what would you do differently?" "What would you do if the circumstances were different?"

3.6

Avoid confusing the other person with multiple questions all at once.

4   Changing Style

In discussions people will have a preference for one of four concepts[15]: ideas, process, people, and action. One of these will be their dominant 'hot button'. There may also be a strong secondary preference, but the others will be weaker. Try to evaluate which orientation switches them on, and modify your interaction accordingly:

---

15 From: *Training for the Cross-Cultural Mind* by Pierre Casse, second edition, Society of Intercultural Education, February 1981.

4.1

Communicating with an idea-oriented person:

a. Allow enough time for discussion.
b. Do not get impatient when they go off on tangents.
c. Be conceptual. Start by expressing the topic under discussion in broader terms.
d. Stress the uniqueness of the idea or topic at hand.
e. Emphasize future value or relate the impact of the idea on the future.
f. If writing to an idea-oriented person, try to stress the key concepts, which underlie your proposal or recommendation right at the outset. Start off with an overall statement and work towards the more particular.

4.2

Communicating with a process-oriented person:

a. State the facts; be precise.
b. Organize your presentation in logical order:
    i. Background
    ii. Present Situation
    iii. Outcome
c. Break down your recommendations.
d. Include options (consider alternatives) with pros and cons.
e. Do not rush a process-oriented person.
f. Outline your proposal in sequence (1, 2, 3 …)

4.3

Communicating with a people-oriented person:

a. Do not start the discussion right away. Allow for small talk. Use an informal style, and try to build a rapport before your main argument.
b. Stress the associations between your proposal and the people concerned.
c. Show how the idea worked well in the past.
d. Indicate support from well-respected people.
e. Use an informal writing style.

4.4

Communicating with an action-oriented person:

a. State the conclusion right at the start, and focus on the results first of all.
b. State your best recommendation. Do not offer many alternatives.
c. Be as brief as possible.
d. Emphasize the practicality of your ideas.
e. Use visual aids.

5   Team Role Style

Listed below are some phrases that are representative of the behaviour associated with the Belbin team roles. See Part V, Section R Team Roles. Use similar kinds of expression when dealing with people you think have the dominant role behaviour. They may well relate more effectively to what you want to achieve.

5.1

To a Chairman/coordinator:

a. "I think what you are trying to achieve is ...."
b. "Let's get back to the main issue."
c. "Let's keep the objective in sight."
d. "To summarize, the key points seem to be ...."

5.2

To a Shaper:

a. "Can you see if you can get things moving?"
b. "The most important issue is ...."
c. "We shouldn't waste time, let's get on with it."
d. "What needs doing is ...."

5.3

To a Plant:

a. "What do you think about this approach?"
b. "Looking at it from a different angle ...."
c. "We mustn't forget the input from ...."
d. "Perhaps you could remind us of the basics?"

5.4

To a Resource Investigator:

a. "I have a contact you could use."
b. "That's a good idea, how can ...? "
c. "Do you know anyone who can ...?"
d. "Can you see if you can get ...?"

5.5

To a Team Worker:

a. "Do you think that you could sort out ...?"
b. "Why don't you help resolve the conflict with ...?"
c. "Would you act as the umpire during ...?"
d. "Can you check that everyone agrees with ...?"

5.6

To an Implementer/company worker:

a. "Would you get into the detail and check that …?"
b. "Can you see if it can be done?"
c. "Why don't you write the information up on the board?"
d. "Do you think that you could provide a detailed analysis by …?"

5.7

To a Monitor Evaluator:

a. "Will you keep an eye out for …?"
b. "What's the problem with …?"
c. "Think it over and give me your opinion on the best alternative."
d. "Would you evaluate all the schedule options available?"

5.8

Completer Finisher:

a. "Check that all the forms have been filled in correctly."
b. "Please give your undivided attention to …."
c. "Please check the 'small print' in the specification."
d. "Make sure that all the activities are complete before we …."

6 Finalizing the Interaction

6.1

At the end of any interaction with other people, make sure that all the points have been discussed to the satisfaction of both parties. Summarize the key points.

6.2

Agree what actions are necessary and who is responsible for seeing them carried out. Allocate a time for when the actions need to be completed.

6.3

Do other people/management need to be informed?

7 Giving and Receiving Feedback

'Faults are more easily recognised in the works of others than in our own.'
   Leonardo Da Vinci's *Thoughts on Art and Life.*

7.1 Giving Feedback

7.1.1
Avoid personal comments and focus on behaviour. Describe what the person did and what you saw happening. *Say how it affected you.*

7.1.2
Choose specific examples (limit the number) that will help the receiver change their behaviour.

7.1.3
An effective formula after presentations is: two positive statements and one 'wish' (for behaviour change).

7.1.4
Don't take someone apart in a feedback session unless you can put them back together again.

7.2 Receiving Feedback

7.2.1
Receiving feedback can be difficult or even painful. Take it on the chin, that is: with an open mind. Whether you agree with the comments or not, the comments should reflect the giver's perception of reality.

7.2.2
Feedback will be a reflection on how your message has been received, and you may have to modify your style if you want to influence people.

7.2.3
Don't be defensive and try and explain your actions. If necessary, ask questions to better understand the information. Ask for suggestions for how you could have done things differently. Ask others what they were thinking.

7.2.4
The more difficult part is doing something about the feedback you receive. Listening to it and making use of it is one of the most effective forms of self-improvement. It's valuable stuff!

8 Dealing with Difficult People

When someone demonstrates a difficult attitude, it is usually because they want to get a reaction from you. If you do, it gives them satisfaction and encourages them. So don't react. Nevertheless, the situation needs to be addressed; it will only get worse if it is not tackled.

It is generally agreed that the following is the way to deal with difficult people:

8.1

Firstly, do a mental check. Are they really being difficult or are you over-reacting? The less reactive you are, the more you will be able to concentrate on solving the problem.

8.2

I find that controlling my 'fight' response, in order to win the argument, is the most difficult aspect of dealing with difficult people. We need to learn this skill; otherwise all the other techniques will fail.

8.2.1

We need to confront the difficult person whilst controlling our emotions and maintaining some objectivity. Try not to take matters personally.

8.2.2

Take a moment to think and consider the other person's perspective.

8.3

Sometimes a person is difficult because you have misunderstood what they are trying to say. Under these circumstances, try and use someone else (the team worker perhaps) to unravel the problem.

8.4

Difficult people tend to take the offensive and say what, when, and why they can't do things. Respond with questions of: what can you/when can you/why can't you?

8.5

If you have to respond to an irrational attack, at say a meeting, then ask the person why they are upset. This demonstrates a willingness to communicate and puts the onus on the other person to explain themselves.

8.5.1

Avoid accusations; turn things around and express how things impacted on you. Use 'I' sentences to explain your experiences. A powerful tool is to explain how you feel as a result of what has been said. How you feel can't be denied by the other person.

8.5.2

Use a negotiating technique and separate the people issues from the problem. If you have an objection to a particular course of action or idea, make sure that it is the issue that is the problem and not the person expressing it.

8.6

If the difficulty is occurring outside a meeting, ask the person concerned for a private discussion. Keep your cool and be pleasant and polite.

8.7

If the difficult behaviour is ongoing, discuss your experience with a *trusted* colleague, such as another project manager, preferably one who had the same difficult person on their project team and find out how they managed matters.

8.8

When receiving a difficult letter/e-mail/memo, it is much easier to restrain an emotional response. Adopt a rule to leave the item in your 'in tray' for twenty-four hours before deciding how to answer it. Write the response and, if necessary, wait another twenty-four hours and re-read your response. No harm can be done by waiting, but an immediate emotional reaction to defend your position will probably make matters worse.

8.9

If you have the talent to be funny, use humour. It defuses aggression.

9   Being Angry

9.1

'Anyone can become angry – that is easy. But to be angry with the right person, to the right degree, at the right time, for the right purpose, and in the right way – this is not easy.'[16]

9.2

If someone is angry with you, let them 'have a go'. *Listen* and let them 'talk themselves out'. Then:

a. If appropriate, apologize for misunderstanding them, followed by: "Let me see if I have understood the problem correctly." If they don't agree with your understanding, ask them to restate their concerns.
b. Stay calm and talk more slowly. Separate the people from the problem.
c. It can help if they have a colleague with them who can defuse the emotions.
d. It can also help if you can use humour.

---

16 Aristotle, *The Nicomachean Ethics*.

10   Priorities

10.1

By defining your own priorities, you will make sure that you will not be fulfilling other people's priorities at the expense of your own. Sort tasks into:

a. Active and reactive
b. Urgent and important

10.1.1
Active tasks are usually what your job is about and tend to be done on your own initiative. They are strategic, positive, and long term.

10.1.2
Reactive tasks are the tactical junk generated by other people. They are short term and will get done by others in your absence.

10.2

For urgent and important, see Section H Prioritising Techniques, subsection 2 Graphical Plots.

11   Time Management

11.1

See Section B, Leadership and Motivation, paragraph 3.3.

11.2

Don't overlook the time management aspects of subsection 10 and paragraph 1.3, both above.

11.3

If you touch a piece of paper, there are only three things you can do with it:

a. Action it
b. File it
c. Bin it

11.3.1
Do not put it back where you found it or shuffle it around your desk because it looks too difficult to deal with. Putting it back to deal with later is a time waster; you will only have to pick it up again and do what you should have done in the first place.

11.4

Don't get sucked into other people's problems.

11.5

Avoid the pitfalls of multi-tasking. Every time you change from one task to the next, there is a catching-up process, a relearning curve, or a relaunch. As a result, everything takes longer.

11.6

Do the formatting for all word processing tasks – letters, reports and documents, last of all the activities associated with their production.

12   Learning

Argyris[17] argues that professionals are often very bad at what he has called 'double-loop learning'. Double loop learning can be described as the ability not only to solve problems by using existing knowledge/procedures ('single loop learning'), but also to innovate and critically reflect on their own behaviour, change it continuously, and thereby adapt to the current situation. Argyris suggests that because many professionals are very good at what they do, they rarely experience failure. For that reason they don't know how to learn from failure. Argyris says: 'Whenever their single-loop learning strategies go wrong they become defensive, screen out criticism, and put the "blame" on anyone and everyone but themselves. In short, their ability to learn shuts down precisely at the moment they need it the most.' Argyris thus claims that professionals have a propensity to behave defensively.

12.1

Continuing professional development (CPD) is a fundamental part of being a member of any of the major professions. It is the systematic acquisition of knowledge and skills throughout your career. It assures your competence and enables you to move jobs more easily and supports longer-term career development. It enhances your overall professionalism. So, think about it, plan it, and do it! It won't be long before it's mandatory.

a. Plan your needs and take ownership of your own learning and development.
b. The easy CPD activities are: to go on a training course, attend a lecture, read *Project* magazine, read other magazines relevant to project management, or more difficult, prepare a paper for publication, and so on.
c. Record what you have done to meet these needs in a reasonable time after the activity. Form a habit so that it becomes a routine when you perform other administrative activities.
d. Consider the learning outcome of the activity and identify further needs.

---

17 Adapted from writings of Chris Argyris and Donald Schön om double loop learning, 1974. Also see Chris Argyris, 'Teaching Smart People How to Learn'. *Harvard Business Review*. 1991

12.2

A reminder:

Tell me – I'll forget.
Show me – I'll remember.
Involve me – I'll understand.

13   Motivating Skills

See Section B Leadership and Motivation, subsection 4. Task, Team, Individual.

14   Some Personal Advice

14.1

Look the part. Buy a decent suit. If you are not a standard size or shape, have it tailor-made. It will be a worthwhile investment. Make sure your shoes are clean. You never know who will be influenced by such detail.

14.2

Have two or three professional photographs taken for those key publication events.

14.3

You never know when you might be called upon to 'say a few words.' Have a speech ready for every formal and semi-formal occasion that you attend Have a one-minute speech for that invaluable occasion when you are in the lift (on your own) with the chief executive or other senior manager.

14.4

Write the occasional article for a technical/professional magazine. Tell colleagues and selected other experts what you are trying to write and ask for their views. This will produce something that might get noticed. It looks good on a CV and is excellent PR for your career.

14.5

Work at your presentation skills.

14.6

Inject a 'smile' into your written and oral communication.

14.7

Find and develop a confidant(e). A person you can trust and share your secrets and feelings with. You must be able to let off steam to someone: a peer or colleague, someone you can relate to, feel comfortable talking to and, above all, who will keep your secrets.

15   Questionnaires

If you wish to understand more about yourself at a more in-depth level, then it is recommended that you talk to your training manager about completing courses or questionnaires that address the following:

15.1

Myers Briggs MBTI Test: This is a personality profile developed from the work of Carl G Jung. It measures four dimensions:

- How we orientate to the world
- How we perceive the world
- How we make decisions
- Lifestyle attitudes

– Extroverted or introverted
– Sensing or intuition
– Thinking or feeling
– Judgement or perception

15.2

Thomas Kilmann Conflict Mode Instrument: The TKI identifies five different styles of conflict: competing (assertive, uncooperative) avoiding (unassertive, uncooperative), accommodating (unassertive, cooperative), collaborating (assertive, cooperative) and compromising (intermediate assertiveness and cooperativeness).

15.3

Belbin Self Perception Inventory of Team Role Theory: You will need to complete this inventory through the Belbin web site. The eight team roles are described in Part V, Section Q.

There is plenty of reference material on the Internet for all three of the above highly recommended questionnaires.

Section F   Politics in Projects

"Politics is not an exact science …but an art. The art of the possible."
Composite from speeches by Otto Von Bismarck, 1863 and 1884

"Politics are almost as exciting as war, and quite as dangerous. In war you can only be killed once, but in politics many times."
Remark by Sir Winston Churchill, 1920

Like it or not, as a project manager, you will be involved in politics, and whatever your view, you cannot avoid it. Politics will be far more prevalent in a multi-project business environment than in companies in the technological industries with strong project cultures. This section tries to address both environments.

Here are a few ways of looking at politics:

- The art of winning (the argument)
- The art of getting things done through people
- How to block or stop things getting done
- The pursuit of power
- A waste of my time
- Using different forms of power to get your way
- Playing the game
- Avoiding confronting difficult issues
- Getting people with different and hidden agendas to agree
- Understanding and influencing key players to bring them around to your way of thinking

There are two ways of approaching politics on a project. If the objective is to meet one's own need at the expense of another, then you are likely to be engaged in political manipulation. If, on the other hand, you are attempting to meet project needs, as well as the other person's need, you are likely to be engaged in influencing.

Effective project managers develop their political skills and influence in organizations. Without authoritative power, the project manager must find other ways to persuade, cajole, buy-in, and 'kick' others to achieve their objectives. Examples of this include:

- Understanding who the key players are and who makes the decisions
- Knowing what switches key players on and off
- Recognising which projects carry sufficient power and profile to stand a chance of succeeding and getting the right resources
- Choosing which battles to fight and when to go with the flow or, as a colleague and I used to say to each other, when to lower your periscope and 'run silent, run deep'.
- Being seen achieving
- Being involved in achieving corporate objectives
- Making a point at the right time
- Recognising who can delay or block a project
- Building trust by helping others to win
- Building alliances with functional managers in order to receive favourable treatment in the allocation of resources

- Building a strong alliance with a senior manager but also some back-up alliances for when the senior manager loses favour
- Understanding the characteristics of the different cultural units

In order to determine the political forces affecting your project, one of your earliest actions should be to plot the power and influence of all stakeholders against their attitude. A stakeholder is anyone who interacts with the project in any form. You need to decide who you need to influence to make your project succeed. Plot each individual within a grid using 1 to 10 scales on both the X and Y axes. See Figure VI.F.1.

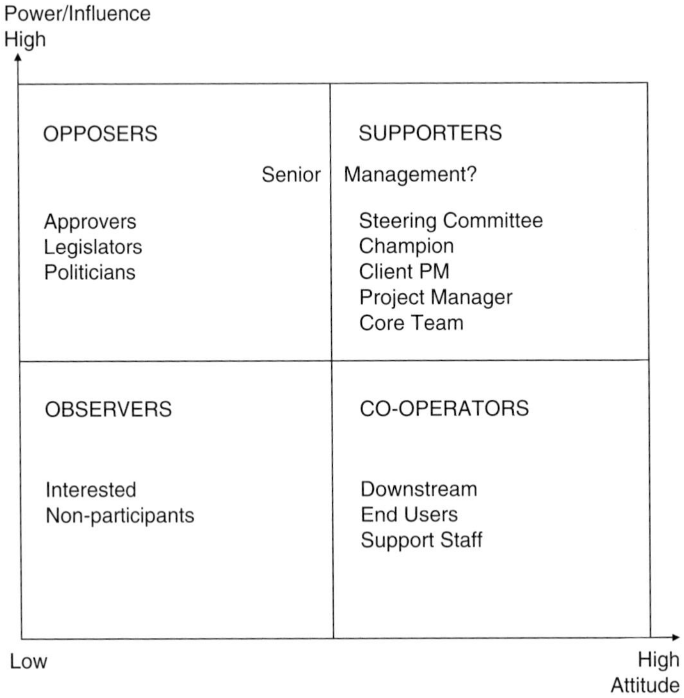

Power/Influence
High

OPPOSERS	Senior	SUPPORTERS Management?
Approvers Legislators Politicians		Steering Committee Champion Client PM Project Manager Core Team
OBSERVERS		CO-OPERATORS
Interested Non-participants		Downstream End Users Support Staff

Low   High
                                                                          Attitude

Figure VI.F.1

All of senior management should be supporters. However, they have been shown as partly in a negative role, since it is entirely possible for an internal business project not to have the support of all the senior management team.

Perform a NICE analysis: What do they *need*? What *interests* them? What are they *concerned* about? What do they *expect* (from us)? Plans then need to be developed for each stakeholder who needs to be persuaded.

The opposers need to be kept informed. Monitor the observers. Involve the cooperators, and leverage the supporters.

1   Typical Destructive Behaviour

a. Lack of agreement on the need for the project or its objectives
b. Attempts to steer projects to meet one individual's needs above those of the whole project

   c. Fights over resources

   d. Lip service paid to decisions but individual actions not implemented. How often have you left a meeting, saying to a colleague in *'sotto voce'*: "Well, that was a waste of time"?

   e. Lack of decisions

   f. Disowning problems and solutions

   g. Blocking of suggestions and ideas

   h. Inflated budgets and deadlines

   i. Offering unsuitable staff for secondment onto projects

   j. Blocking ideas and hard plans to escape accountability

   k. Delaying projects through a lack of decisions

   l. Looking for a scapegoat when things go wrong

   m. Claiming credit when things go right

   n. Concealing or withholding information

2   Dubious Behaviour?

2.1

Canvassing project team members and decision takers to get them to agree

2.2

Consulting others to get them to buy into the project's goals

2.3

Spending time and socializing with people who have power and influence

2.4

When decisions have to be made and you are providing options for the boss, always include one that is easy/obvious (not too obvious!) for them to reject.

2.5

Copying memos to your manager's boss

2.6

If you have to have an argument with the client, try doing it through someone else. For example, use your project controls manager to argue with their opposite number.

2.7

Managing the client's expectations to achieve Success: $S=D/E$, (deliverable/expectation). At the start of the project, tell the client how tough the targets are that they have set. Complain that the changes will delay the project. Tell them all your problems. Manage their expectations downwards. At the end of the project, give the key users something that is important to them; do a bit more than usual for the client. Increase the deliverables where possible, provided it costs only a little money.

2.8

Using 'clever' language. For example: one construction manager when asked: "How are things?" always said: "Fantastic." He never said if they were fantastically good or fantastically bad!

Reasonable is a word that is usually interpreted by the receiver within their own mental context and, consequently, the issue being discussed is received favourably.

2.8.1
Be wary of people using phrases that usually mean the opposite of what is actually said:

a. I wouldn't worry if I were you; this is no criticism of you personally.
b. I'm sure nobody will notice.
c. It's no problem at all.
d. I'm sure you're right.
e. I'd love to help you if I could.
f. It's really very simple.
g. I'll attend if I possibly can.

3   How Politics Can Affect a Project

Lack of political awareness on projects can cause problems in each stage of the project life cycle. For example:

3.1

Terms of Reference:

a. No agreement to project objectives
b. Wrong people seconded onto the project team
c. Unrealistic deadlines or constraints; for example the budget

3.2

Feasibility Study:

a. Favouring an individual's preferred solution
b. Blocking new and different ideas
c. Constantly changing the objectives

3.3

Planning:

a. Artificially extending deadlines
b. Overestimating resource requirements
c. Planning and re-planning to fit individual steering group members' wants, irrespective of whether this achieves anything

3.4

Developing, Executing, and Setting to Work:

a. Avoiding taking on actions, always delegating them to the rest of the team
b. Using the operational position as an excuse for lack of delivery on project actions
c. Changing the objectives and priorities
d. Going back on decisions

3.5 Evaluating/Post Project Appraisal:

a. Making things look better than they were
b. Criticising all of the things that went wrong without recognising the things that went well
c. Not evaluating projects to avoid confronting poor performance or results

4 Some Advice

4.1

As Churchill said: "In politics you don't have time. You only have moments." Pick your moments carefully.

4.2

Be bold, make decisions, and do things that achieve results. Remember that it is much easier to apologise than to ask for permission.

4.3

Know the company procedures well, not necessarily because you are going to use them, but because you are going to use the cracks between them to achieve what you want to do.

4.4

Don't take on a more senior person unless you are prepared to go for broke and accept that you may have to resign. Remember senior management will always support the supervisor (as a project manager, this may sometimes put you in a stronger position – depending on the company's project management culture).

4.5

Remember there are always at least three sides to a story.

4.6

If you lack knowledge of a particular subject area, back the proposition from the team member who has the most logical and clearly presented argument. Choose the simplest proposition or solution.[18]

4.7

Never get the contract out in front of the client.

4.8

If you want to get promoted or move onto a better assignment, start looking for your replacement early and train them.

4.9

Train all your people through the Blanchard management cycle. See Section B Leadership and Motivation, subsection 3.

4.9.1

Nominate a different person from your direct reports (unless you have a deputy project manager) to take over for you every time you have to be absent from the project.

4.10

If, as part of your MBWA (see Section B, subsection 6), you talk directly to junior team members always stop off and tell their supervisor what it was all about. However, before wandering into their domain it is sensible to advise the supervisor of a group that that is what you propose to do.

4.11

Have a regular updating meeting with your boss; keep them informed. Before leaving, be sure that they have bought in and agreed to your proposals.

4.12

Train people to keep written records of conversations with the client and always report them to you, the project manager.

---

18 'Occham's (or Ockham's) Razor' (for separating competing hypotheses) is interpreted to mean that when you have two competing theories that make exactly the same predictions, the simpler one is the better one. With competing propositions, the one with the fewest assumptions should be selected. The fewer assumptions that are made, the better.

4.13

Don't forget to have your 'lift speech' ready for a senior manager that you want to influence. See also Section E Personal Skills, paragraph 14.3

4.14

CYA. Keep a project journal.

5   Something to Think About

5.1

'The squeaky wheel gets the oil.'

5.2

You may think that the route to success is to keep your head down, work hard, overcome problems, and deliver competent work. This won't necessarily get you promoted. The reward for hard work is more work, rather than appreciation! You may even find colleagues offloading some of their work onto you because you are 'obviously the right person to do it, and you are so much better at doing it than we are.'

5.2.1
You need to make sure management are aware of your achievements. It is hard not to respond positively to someone's enthusiasm at their achievements. If you want to get on, get noticed.

5.2.2
Those who keep the business running tend to be left alone to get on with it. The ones who move up the organization are the ones who boast about their successes and who demand more out of their jobs.

Section G   Presentation Skills

"The ability to communicate effectively through the spoken word is essential to success in business management."

Henry Ford II

You should volunteer for every opportunity to make presentations, both internally and in the public arena. A presentation gives you an opportunity to demonstrate your experience, knowledge, and capability. It can influence senior managers, impress colleagues and lift your career to a new level. It can enable you to be considered for assignments that otherwise may not have been open to you. It can also open up opportunities in the market place. In presenting to the project team; it is the project manager's chance to demonstrate their command of the issues and situation.

'All memorable speeches are known by a few words, you have to be ready with a singular theme, ("I have a dream", "tryst with destiny", "ask not what your country can do for you"). The pithy formula that encapsulates the address, the soundbite, is not a craven capitulation to mainstream media, it is the organizing principle of the speech. "To be or not to be. That is the question." It is the message caught in a phrase.'[19]

All presentations have the same overall structure, whether Act I, Act II, Act III or (1) tell the people what you are going to tell them; (2) say it; and (3) summarize what you have told them. Specifically, a structure might be: introduction, development, and conclusion or explanation, complication, and recommendation.

In summary, the one golden rule is: Tell people what you are going to tell them, tell them, and tell them what you have told them.

Remember the statistics of communication (the statistics may vary slightly but the proportions are right); people receive the message as follows:

Vocal	The words	7%
Vocal voice	Pace and volume	38%
Body language		55%

1   Fundamentals for All Presentations

1.1

Preparation

a. Some early preparation issues are:
   i. How long do you have to make your presentation? Does this include any discussion time?
   ii. What is the room like, and what are the seating arrangements? Are you familiar with it? See subsection 8 below.
   iii. What are the projection facilities, and what other visual aids can be used? Is the equipment likely to be reliable? Also see subsection 8 below.

---

19 Philip Collins in an article in *The Times*, 25 September 2015, concerning 'the speech [a politician] should give.'

   iv. Prepare a set of notes on filing cards.

b. Remember to tell people the location and title of the talk when you invite them to attend.

c. Do not start while latecomers are still coming in. Wait for the audience to settle down.

1.2   Delivery

a. In your introduction you should explain what your objective is, what you are going to cover, how the presentation will be structured, and how long it will last.

b. It is important to capture your audience's attention with your first few words. Wow them! Memorise your opening statements.

c. Make sure that you can be heard. Project your voice to the back of the room, but don't shout. Vary the pitch, tone, and speed of delivery.

d. Speak with conviction and be *enthusiastic* about your subject. Train yourself not to use 'er', 'um' or 'you know'.

e. Look smart and professional – polish your shoes. Stand upright, shoulders back and feet slightly apart. Avoid distracting mannerisms, particularly jingling coins in your pocket. Remove them.

f. Avoid putting your hands in your pockets!

g. Be still, do not rock back and forward, or shuffle. Nevertheless, you can walk about.

h. Look at individuals in the audience. Vary your pattern of eye contact. Do not do it in sequence. Don't forget eye contact with the wing people on your extreme left and right who are just in/out of your peripheral vision.

i. Do not block the audience's view of the projection screen. Do not admire your visuals by looking at the projection screen with your back to the audience.

j. Be very careful with pointers. A pointer can wave around like a conductor's baton. Rest the pointer against the word or image you want to emphasize on the screen. Avoid laser pointers; they magnify every nervous shake of the hand.

k. Avoid the use of technical terms, jargon, and abbreviations; use layman's language. Words can be misinterpreted. People interpret words within their own business context – it may not be the same as yours.

l. When reading, you can go back and read something again. The audience cannot do this. Consequently, you will need to repeat important points. However, repeat these key issues in a different manner.

m. One of the most powerful techniques in presenting is to pause – particularly after important points.

n. Be yourself and smile. If you can, use humour occasionally.

o. Never apologise for your material. Apologies lead the audience to expect the worst. A no-no, as an example, is: 'I know you can't see this but …'

p. Use the rule of three: "Friends, Romans, Countrymen;" "Blood, sweat, and tears;" "The right materials, in the right quantities, at the right price."

q. Tell a story. People like stories and, most importantly, people remember stories and the point being made with them. Paint a picture with words. Use examples.

r. Relegate detail to supporting documentation as hand-outs.

s. Use rhetorical questions. It makes people sit up and listen if they think they may have to answer a question.

t. If you ad lib or digress, it will cost you time that you are unlikely to make up, unless you leave something out. As a colleague once said to me: "All my ad libs are very carefully scripted."

u. Consider how you will deal with audience participation. Would exercises or case studies be appropriate? See subsection 10 below.

v. It is essential to be properly rehearsed. Rehearsal will enable you to judge how long the presentation will take. This is likely to be an underestimate. Remember the 5 P's: Proper Preparation Prevents Poor Performance.

w. If you use a script for a speech/presentation, underline a couple of words where you will be looking at your audience. This will help you find your place when you look down at your notes again.

x. Remember that presentations are most hazardous when starting or finishing. Memorise your closing sentences and finish on a positive note.

y. Keep to your declared or allocated time. It is better to finish slightly early rather than late.

z. Finally, it is self-evident that you must know your material; otherwise, you will not be able to deal with something that goes wrong.

2   Format for a Presentation to Inform/Explain[20]

2.1   Introduction

a. Establish a rapport, a common bond, or an agreement with the audience. Try one of the following:

   i. If you are to be introduced, talk to the person doing the introduction and indicate what you would like them to say and what to leave out that you will be covering. Get an idea of what else they will say and use some material in the transition to your presentation.

   ii. Use a humorous incident that relates to the audience, the occasion, or theme of your presentation.

   iii. Refer to the importance of the occasion.

   iv. Give your audience recognition for some specific achievement.

b. Briefly describe the subject you will be covering, supported by a visual aid.

c. Explain why they need to listen. Tell the audience in one or two sentences (support with a visual aid) why the information you will be revealing is important, interesting, and will be useful to them.

d. Provide an outline framework (support with a visual aid) explaining the *what* and *how*.

2.2   Body of the Talk

a. Use one of the basic approaches:

   i. Order of importance

   ii. Form of logic

   iii. Deductive: general to specific; or inductive: specific to general

---

20 Subsections 2, 3 and 4 are extracts from a Cranfield lecture handout.

    iv. Cause to effect or effect to cause

    v. Time or sequence

    vi. Question format

b. Simplify the task of following the development of your subject.

    i. Number the points so that the audience can see progress.

    ii. Use examples, facts, and statistics supported with visuals to show trends and relationships and highlight facts.

    iii. Use forms of evidence, such as, personal experience, judgement of experts, or analogies.

2.3 Conclusion

a. Summarize in reverse order of importance.

b. Emphasize where the information will be useful and helpful to the audience.

3 Presentation to Influence/Convince

3.1 Introduction

a. Establish a rapport or a common bond with the audience:

    i. Refer to how you were introduced.

    ii. Relieve tension.

    iii. Acknowledge that it is a privilege to be there.

    iv. Refer to the importance of the occasion.

    v. Give your audience recognition for some specific achievement.

b. Define the purpose of the presentation with a visual aid:

    i. Provide a statement of the subject, premise, theme, or point of view.

    ii. Define the scope of the talk.

    iii. Explain the style of your presentation.

c. Establish the importance, timeliness, or impact of the subject (support with a visual aid).

    i. Explain the benefits to be had by adopting, agreeing to, or acting upon the recommendations.

    ii. Emphasize the advantages for the individuals present.

    iii. Quote a startling or arresting fact.

3.2 Body of the Talk

a. State the findings and conclusions using, where possible, each of the following, and support each one with a visual aid:

    i. State the statistics and/or facts in support of findings.

    ii. Use a personal experience.

    iii. Use an analogy to explain the conclusions.

    iv. Use examples to illustrate the findings.

    v. Explain where experts have come to similar conclusions.

b. State the recommendations.

3.3 Conclusions

a. Summarize the evidence and conclusions.
b. Summarize the recommendations and the major benefits that will be helpful to the audience (support with a visual aid).
c. Request action to be taken (support with a visual aid).

4 Presentation Expressing a Viewpoint/Opinion

4.1 Introduction

a. State what the subject is and why you are going to talk about it.
b. Express your point of view and where you stand on the subject.

4.2 Body of the Talk

a. Support your point of view with a personal experience.
    i. What event happened?
    ii. When did it occur?
    iii. Who was involved?
    iv. Why did it happen, and what was the cause or effect?
b. Provide an analogy.
c. What the judgement of the experts is.
    i. Who the experts are
    ii. What credentials they have
    iii. What the experts said
d. Use examples from research and experience.
e. Provide statistics and facts supported by research.

4.3 Conclusions

a. Restate your point of view.
b. Recommend what action should be taken.

5 Team Presentations

See Part III, Section E Tendering and Proposal Phase, subsection 9 for additional details of presenting a proposal or tender to a client. For completeness, this subsection 5 is a summary of the issues amplified in Part III, Section E.

All of the issues raised in the other paragraphs of this section apply; however, there are some important differences and additional problems with team presentations.

5.1

The team leader has two roles: to manage the agenda for the whole group, allocating roles and responsibilities for the tasks that have to be done and, in addition, they have to prepare and deliver their own presentation.

5.2

In a presentation of this nature, it will be helpful to have name place cards.

5.3

Use a team logo and ensure there is consistency in the design of the visual aids.

5.4

Identify a theme for the presentation that links every individual presentation.

5.5

An issue to be addressed is how to hand over from one speaker to the next. Also, it may sound obvious, but make sure you know each other's names!

5.6

With a number of individual presentations, there are bound to be interruptions or questions; consequently, you will have timing problems. It is, therefore, essential to have identified one or two portions that can be omitted.

5.7

Eliminate areas of overlapping content.

5.8

Look interested in other speakers.

5.9

The team leader should take the responsibility of accepting all questions and then allocating them to a particular member of the team.

5.10

This type of presentation needs additional team rehearsals in addition to the individual preparation. Use a red team review process. See Part III, Section E, paragraph 9.19.

6   Your Audience

6.1

Everything should be presented in terms of the audience's interests or point of view. Consequently, find out how many people will be in the audience, what experience they have, why they are interested, and what companies they work for.

6.2

When preparing your material you need to consider how much your audience knows about the subject matter. With a large group, you should assume that some people will know as much as you about certain aspects of your subject. This will help you decide on how much detail you need to go into and what you can leave out.

6.3

As part of your preparation, you should also consider the composition of your audience. What type of people are they? What 'switches them on', and what appeals to them? As indicated in Section E, Personal Skills, subsection 4, are they people who are fascinated by ideas, by the processes, or by action or are they concerned for people issues? They will be a mixture, but which is their stronger or dominant interest? Consequently, your presentation needs to use phrases that catch their attention, and change your style accordingly.

6.4

You will also need to consider your audience's attitude. This may depend on how many are in the audience – find out. How might they react to what you propose to say? Are you planning to say anything startling or include anything unexpected? How has the audience been briefed?

6.5

Before the formalities begin, make contact with people and mingle with the audience. Try to develop a relationship with some of them, particularly those who might be hostile to you.

7   Presentation Skills Analysis

7.1   The Introduction

a. Was the subject introduced in an interesting way?
b. Was it clear when the introduction had ended?

7.2   Voice

a. Was there sufficient variation of speed?
b. Was there sufficient variation of volume?
c. Was the voice projected effectively?
d. Were there adequate pauses?
e. Were the words articulated and enunciated clearly?

7.3   Audience Contact

a. Was there sufficient eye contact with the audience?
b. Would *every* member of the audience have understood *every* word? (For example, eliminate jargon.)
c. Were there any irritating mannerisms?
d. Was there sufficient enthusiasm for the subject?

7.4   Speech Content

a. Was there the right amount of information for the time allowed?
b. Did the speech contain things that need not have been included?
c. Were the various facts well balanced?
d. Would more 'word pictures' or analogies have helped?
e. Was the message convincing?

7.5   Visual Aids

a. Were there adequate visual aids, and were they appropriate?
b. Could the composition of the visual aids have been improved?
c. Was contact with the audience maintained during the visual aids?
d. Was there a good balance between the spoken word and the visuals?

7.6   The Conclusion

a. Was it clear when the conclusion had arrived?
b. Was a summary or recap required? If so, was it adequate?
c. Did the conclusion follow logically from what was said?
d. Did the conclusion leave the audience with a message and support the stated objective?

7.7   Overall

a. Did the presenter give the impression of knowing their subject thoroughly?
b. Was the presentation well organized, and did it have a logical structure?
c. Did the presentation grab your attention, and was it interesting?
d. Was the presenter enthusiastic about the subject?
e. Did the presentation finish on time?

8   Organizing the Location

8.1

Make your audience comfortable in the proposed room/location. Select a room with plenty of space and some soft furnishings to improve acoustics and prevent echoes. Extra space will allow you to move around. You may also need extra space for group break-out work. If you are running a training programme, you will need to arrange separate break-out rooms.

Ideally you want a dimmer switch for the main room with a separate light switch for the presentation/screen area – alternatively, remove bulbs directly above or behind the screen.

8.2

If you can, select comfortable chairs. Arrange them so that they are not too cramped. Allow space so that people can move or shift position in their chairs. Check the seating layout for a clear view of the screen and flip charts.

8.3

Check that all equipment is working and make sure that lenses are clean. If you are operating the equipment yourself, get a professional to demonstrate how it all works. Ask what can go wrong and arrange a contact number for emergencies or if spares are needed.

8.4

Have as large a projection screen as possible. Arrange for flip charts (two preferably, one each side of the presentation screen) and whiteboards with associated felt-tipped pens (minimum three colours). Have spare charts available if it is a long session.

8.5

Make sure you know how to operate the heating and/or ventilating equipment. Again, get the professional to run through how to adjust it. You will probably have to change it as the room warms up from body heat.

8.6

Get the power cable taped to the floor. If you disturb it, you might dislodge the power plug without pulling it out of the socket. Your presentation will then continue on the laptop battery, and sometime later your presentation will come to a halt without it being obvious what is wrong!

8.7

Provide name badges and place cards. You will find this particularly helpful t in the early stages of a project when you are trying to remember everyone's names.

8.8

Allow some time for breaks for informal discussions and for toilet visits.

8.9

Make sure that there will be the minimum of interruptions. Insist that mobile telephones are switched off. If you are presenting in an office area, make arrangements to stop the telephones. Arrange for messages to be pinned on a board outside the room and accessed during breaks.

8.10

Minimise distractions; arrange for a clock at the back of the room – to help with your time management. Remove any clock in front of the audience to stop 'clock watching'.

9 Visual and Other Aids

9.1

Visual aids are necessary because people's attention span is about 7.5 minutes. As stated, they should be aids to communication. Keep them simple, bold, and relevant. Have a consistent theme or style to your slides. Numbering slides can sometimes be helpful.

a. Make the content large, simple and bold. Avoid being over artistic with today's software.
b. Convert statistics into charts or graphs.
c. Keep organization charts uncomplicated. Complex ones invariably get misinterpreted. If you cannot avoid a complex situation, build it up in stages.
d. Relegate detail to supporting documentation as hand-outs.
e. Use photographs.

9.2

Talk through the contents of your slide and explain the point you are trying to make. Highlight or enlarge each line in turn so as to focus people's attention.

9.3

Use a blank slide or the 'B' key on your laptop to blank out a slide and prevent people looking at a slide you have finished with.

9.4

Use visuals other than just the projected images. Use a flip chart or whiteboard. Also consider using a book or other document to hold in your hand as the authority you are quoting.

9.5

Use the flip chart from the back to the front. This enables you to turn the pages without having to worry about pushing the paper over the top of the stand. Pull the paper over the top, turn to the audience to explain what you are going to do, and gravity will let the paper fall.

9.6

Write clearly. If you can't write, *print*.

9.7

Prepare your flip chart beforehand by writing on it in pencil. Your audience will then be impressed with how well you remembered the numbers and with your mental arithmetic.

9.8

To stop a page being a distraction, turn over to a new blank page when you have finished with a particular sheet. You do not want people to continue reading what is on the old chart when you have moved on to another subject.

9.9

Use sticky notes or turn over the corners to help find a particular sheet.

9.10

It can be debatable as to whether you should provide a copy of your presentation for the audience to make notes on as you present. Alternately, should you tell the audience that you will provide a copy at the end? If you are dealing with a responsible audience, you should take the risk that they will not read ahead.

If you use a slide with a question, arrange the handout with a blank so that the answer appears on the next page.

Distribute handouts for detailed material at the end.

10   Dealing with Questions

10.1

Your introduction should tell the audience how questions will be handled. With a small group, it will be easier to manage questions as you progress (but watch that you do not compromise the overall timing). With a very large group, you may need a chairman to manage the grouping of questions.

10.2

You can tell people to ask questions at the end, but you may not be able to prevent them. However, if you take questions as you progress, you will demonstrate command of your subject and confidence as a presenter.

10.3

In a seminar format, you can suggest breaking out into discussion groups at the end of your formal presentation.

10.4

Make sure everyone has heard the question by repeating it. Alternatively, ask the questioner to repeat it, thus giving yourself thinking time.

10.5

Write the question on a flip chart/whiteboard if you need additional thinking time.

10.6

Make sure that you have understood the question by paraphrasing it into your own words and check with the questioner: "Is that what you meant?"

10.7

Refer the question to an expert colleague or to a specific member of the audience. Alternatively, throw the question open for general discussion, and ask the audience what they think.

10.8

If you don't know the answer, say so. This can give you credibility. Say you will find out and let them know. Make sure you do.

10.9

If you have allowed for questions at the end and there are none, ask the audience a question. This might help to stimulate the group.

11 Summarizing a Presentation

a. Consider your audience.
b. Decide on your key points.
c. Decide on the format – to inform, to persuade, or to express an opinion.
d. Prepare the room. Prepare your note cards. Prepare your visual aids.
e. Rehearse and get your timing right. Rehearse your visual aids.
f. Set up and check the equipment.
g. Calm down; deep breathing beforehand helps.
h. Smile, even if you are nervous.
i. Hook your audience with your first words.
j. Tell them what you will tell them.
k. Deliver. Tell them.
l. Punch home your message. Tell them what you have told them.
m. Ask for questions.
n. Regain control and leave the stage to applause!

Section H   Prioritising Techniques

This section is about identifying *which task* for the subsequent allocation of *who does it*.

1   Group Work Using Flip Charts

1.1

List the issues on a flip chart and use the experience, knowledge, and judgement of the team to decide on the key issues. Use MoSCoW: *must* haves, *should* haves, *could* haves and *won't* haves. This will probably result in too much discussion or even generate arguments. Consequently, a voting process will be needed for a quicker selection; see 1.2 and 1.3.

1.2

List the issues on a flip chart and use a simple voting process.

a. Identify the decision-making criteria such as 'the most likely cause of the problem' or 'the least cost and shortest schedule' and so on.
b. The number of single votes per person is equal to the number of issues x 0.2. If you think it is desirable to have a greater number of votes, increase the multiplying factor.
c. Each person allocates their votes to the idea or ideas they feel are most likely to succeed. For example, if the participants have four votes, one person may give three votes to idea number 4 and one vote to idea number 7. Add up all the votes, and select the most popular solution.

1.3

Again, list the issues on a flip chart, and use a prioritising voting process:

a. As above, identify the decision-making criteria.
b. Give each participant five votes and ask them to select five choices, giving five points to their most preferred choice, four points to their next preferred choice and so on.
c. In order to save time, ask each participant to mark their choices directly onto the flip chart.
d. All votes are then added up, and those with no points or with only two or three points are eliminated. This should reduce the list to a manageable five to ten items.

2   Graphical Plots

2.1

Use a two-by-two graphical plot of importance versus urgency. See Figure VI.H.1 below.

a. It may seem strange to have high urgency but low importance as a number one priority. However, if you are one of those people who get anxious or whose brain

Figure VI.H.1

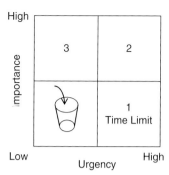

gets befuddled if there are too many issues to deal with at once, then get rid of a lot of them by dealing with them first – but only spend a few minutes on each of them. Limit your time on these items to twenty minutes to half an hour maximum.

b. You will now have a clear head to deal with the ones that matter, the number two priority in the figure.

c. Having got rid of the reactive issues, you can spend some time dealing with the issues that are proactive (number three in figure), the ones that your job is about and help move the project forward.

2.2

Use a two-by-two graphical plot of impact versus difficulty, see Figure VI.H.2.

a. This is fairly self-evident. Deal with the easy ones that will provide the most impact first, then the ones that will provide results but are more difficult to tackle. Finally, if you have time, deal with the easy ones that provide least returns.

b. You could use importance and urgency in conjunction with your number 1's and 2's to help sort out your priorities.

Figure VI.H.2

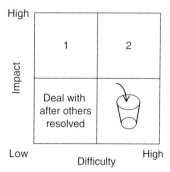

2.3

For prioritising the identified risks, use a two-by-two graphical plot of probability against impact (Figure V.M.2). Alternatively, plot risk (impact x probability) versus the management effort required to resolve the issue (Figure V.M.3).

3 Binary Decision-making

3.1

This is an effective but time-consuming tool for reaching consensus in a group. Everyone is involved in the process, and the process is seen to be fair. Produce a pre-printed form as shown in Figure VI.H.3, and issue a copy to every member of the group. Brainstorm as described in Section K, subsection 1 in order to identify all the issues, solutions or options, and so on. If necessary, use any of the above techniques to reduce the number of issues to a manageable number. Ten should be enough. They should then be written up clearly and neatly on a flip chart for everyone to see.

Preferences

1	1	1	1	1	1	1	1	1
2	3	4	5	6	7	8	9	10
2	2	2	2	2	2	2	2	
3	4	5	6	7	8	9	10	
3	3	3	3	3	3	3		
4	5	6	7	8	9	10		
4	4	4	4	4	4			
5	6	7	8	9	10			
5	5	5	5	5				
6	7	8	9	10				
6	6	6	6					
7	8	9	10					
7	7	7						
8	9	10						
8	8							
9	10							
9								
10								

NumberRank

1	
2	
3	
4	
5	
6	
7	
8	
9	
10	

FigureVI-K-3

Figure VI.H.3

3.2

Each member of the group then makes a whole series of choices. Do they prefer issue 1 or issue 2? They then circle their choice on the first line of the chart. Next, do they prefer issue 1 or issue 3, and then circle their choice. Do they prefer issue 1 or issue 4, and so on?

3.3

The next step is to list how many times number one was circled and then number two and so on. The individual then enters the totals in the numbered column on the

right of the chart (Figure VI.H.3). They are then ranked in the second column in accordance with the number that was preferred most and then the second highest score
and so on.

3.4

Finally, a straightforward mechanistic collection of all the rankings from each member
of the group is taken in order to produce an overall ranking.

Section J   Problem-solving Process

The problem is not that there are problems. The problem is expecting otherwise and thinking that having problems is a problem.

Theodore Rubin

Projects can be described as a series of problems laid end to end. These problems are a series of obstacles preventing the project team from achieving a successful project on time and to budget. Problems are unexpected because if you had expected them, you would have dealt with them. The snag is that the client's reaction is to fix them regardless of the budget available for the activity. It is not an opportunity to redesign the project. Problem-solving is, therefore, an essential skill for the project manager, and it requires one to follow a logical sequence of steps. It may not be necessary to comply with every step in sequence, and following the steps does not guarantee arriving at the best solution. As the project manager, you and the project team may be deciding on a plan of action or recommending one to a client, and by using each step the chances of mistakes are reduced.

In many cases, problems within groups are caused by failure of communication. The project manager must ensure that all project problems are communicated to them regardless of whether the individual involved believes that the problem can be solved or not.

Peter Drucker distinguishes between two types of problems:

The Generic:	These types of problem underlie many situations or occurrences and are solved tactically by a rule, principle, or procedure.
The Exceptional:	This is the unique problem that needs to be solved strategically by special methods, using careful analysis as set out below.

However, if a project problem is of an extremely technical nature, then the preferred solution would be to bring in an expert from a functional group and, in an extreme case, an outside consultant.

The various steps involved in solving the unique problem can be grouped into four broad phases: defining, analysing, selecting, and implementing, as detailed below:

Phase 1	1.	Identify and define the problem.
	2.	Describe the objectives and success criteria.
Phase 2	3.	Analyse the problem by gathering data and finding causes.
Phase 3	4.	Create and propose solutions.
	5.	Evaluate, forecast consequences, and select.
Phase 4	6.	Recommend and plan action.
	7.	Implement solution.
	8	Evaluate outcome and follow-up.

Steps 1 and 2 will help to distinguish between the problem and its symptoms.

It helps to improve one's chances of finding a suitable result by proposing more than one solution in step 4

Steps 5 and 6, forecasting consequences and planning actions for the proposed solution, help to develop less costly implementations.

1   Define the Problem

1.1   Describe the Problem

a. Distinguish between the problem and its symptoms. Assume that all symptoms lie!
b. What is the problem? What does it do? To what degree? Why do it this way? How much does it cost? How long will it take? Who? Where? When?
c. Find the critical factor by asking: "What happens if no action is taken?"
d. Can a cause be stated? What changes took place? What was the unexpected change that caused the deviation?
e. What are the uncontrollable events?

2   Define the Objectives and Success Criteria

2.1

Objectives are:

a. Short term or long term
b. Specific, measurable, achievable, relevant and time related

2.2

Criteria for a successful solution are:

a. Quality, safety, productivity, cost, time, and effect on people.
b. What is the standard or norm that is to be achieved?
c. 'Must haves' are major objectives and decision-making criteria.
d. 'Nice to haves' are wants that can be prioritised.

3   Analyse the Problem

3.1

Understand the problem; what do you need to know? What information is required?

3.2

Gather, Display and Analyse data:

a. What data? What is its relevance or importance? What is the source of the data? What was the time period? Who should collect it?

b. Note any gaps in the data. Are any of the data based on assumptions or guesses?
c. What is the relationship between data elements? Quantify the differences or deviations. Agree on acceptable variations.
d. Establish categories or groupings. Be wary of grouping on the basis of similar symptoms. Look at the variations rather than the problem.
e. Summarize results and present them graphically. This makes the data and their relationships much easier to understand.

3.3

Find and verify causes:
a. Ask questions in order to gather additional data.
b. What are the interactions between the data?
c. Challenge assumptions.
d. What differences are there and what alternatives or similarities?

4   Create and Propose Solutions

4.1

Brainstorm and list ideas – the more the better.

4.2

What can be changed?

4.3

What are alternatives? Disagreement within the team can be useful since it generates the necessary alternatives.

4.4

Combine and improve on ideas.

4.5

Look at things from a different perspective.

5   Evaluate, Forecast Consequences, and Select

5.1

List solutions against the success criteria in 2.2 above.

5.2

Quantify, grade, and rank ideas. How long will they take to implement?

5.3

Forecast the consequences of any actions:

a. What might happen?
b. How easy will it be to reverse the proposed course of action?
c. What impact will the proposed solution have on other activities?

5.4

Identify criteria for selecting the best solution and select it.

6   Recommend, Plan Action, and Implement the Solution

6.1

Recommend the solution to management. Of course, if you have the authority, you will not have to go through this step. However, some of the activities in this step, for example, b and c below are still valid in the process of developing an action plan.

a. Prepare and rehearse for the presentation.
b. Anticipate questions and concerns.
c. Identify advantages and benefits.
d. Obtain approval.

6.2   Develop an Action Plan

a. Write out how the plan will be executed.
b. Identify who will have responsibility for it.
c. Set targets, milestones, and time scales.
d. Identify what support will be required.
e. Evaluate the cost and payback.

6.3

If necessary, revise the action plan.

6.4

Implement the plan.

6.5

Take action.

7   Evaluate the Outcome and Follow Up

7.1

Evaluate the outcome against the success criteria.

a. Check that the solution has solved the problem.
b. Are the anticipated results being achieved?

7.2

Follow up

a. Take corrective action.
b. Critique the experience for future use.
c. Formulate any new problem!

Section K   Problem-solving Techniques

"We can't solve problems by using the same kind of thinking we used when we created them."

Albert Einstein

Problem-solving is principally a process of applying a series of techniques with varying degrees of complexity. These techniques, which can either be used singly or together, are listed below.

All of the techniques rely on the principle that a group is more likely to produce an effective solution than individuals working on their own. Thus, effective teamwork is an essential element of the problem-solving process.

The more the initial problem concerns people, the more important it will be to arrive at a consensus solution.

The solution to many problems is likely to involve change, and in these circumstances, it is even more important to involve those concerned or are affected by the change.

The following techniques are listed roughly in order of effort required for successful implementation:

The following list of techniques includes those generally accepted as problem-solving techniques. However, my experience is that the charts and graphs are more commonly used in reports than as problem-solving tools. Further, whilst the concept of cause and effect sounds useful, I find that it is little used. The cause-and-effect diagram is probably best suited to a manufacturing environment or possibly to the commissioning phase of a process project. Nevertheless, it has been included for completeness.

Apart from brainstorming, the most useful project problem-solving tool is *force field analysis*.

1   Brainstorming[21]

1.1

Brainstorming is a technique that is most needed when there is a deficiency in the *plant* role in the group/team (see Part V, Section R Team Roles, paragraph 1.3). It is perhaps the commonest technique used and is also the most self-evident. Nevertheless, there are a few basic rules that must be followed if it is to be truly effective. Try to limit the number of people present to a manageable group – probably no more than eight to ten.

a. The problem or issue to be addressed, together with the objective of the session, should be written up on a board or flip chart so that it is clear for everyone to see.
b. Allow a few minutes for people to get their thoughts together.

---

21 Edward De Bono, *Six Thinking Hats,* Penguin Books, 1990 & 2016. ISBN 9780140296662. Also http://www.debonogroup.com/six_thinking_hats.php

c. Generate as large a number of ideas as possible. Build on each other's ideas.

d. Record and number all ideas and keep them visible. An appointed scribe is not allowed to change any of the words that have been suggested.

e. Everyone must be given an equal chance to participate in an unrestricted manner. This can be carried out in two ways:

    i. The structured approach is where everyone contributes an idea in turn. A 'pass' is acceptable if no idea comes easily to mind. This method encourages everyone to participate and stops a few people from dominating the process. It can, however, create pressure to contribute.

    ii. An unstructured and more relaxed approach allows everyone to contribute as ideas come to mind. However, it risks domination by the most vocal team members.

    iii. A combination of both methods may generate the most ideas. Use a structured approach until ideas start to dry up, and then use an unstructured process to capture all the remaining ideas.

f. No evaluation of ideas is carried out at this stage since it will slow down the process.

g. Most importantly of all: no criticism is allowed of other people's ideas.

1.2

After a list of ideas has been generated, the team should select some of the most significant ideas by using the following steps.

a. Make sure everyone understands each idea on the list in the same way.

b. Eliminate duplicates.

c. Combine items of a similar nature in a manner that the authors of the original items can agree to.

d. Select the top ideas using a voting system (see Section H Prioritising Techniques).

2   Check Sheets

2.1

Frequently, it is necessary to collect data to enable the team to analyse a problem. A check sheet is a form for recording data. However, before designing the check sheet, it is necessary to determine the following.

a. Firstly, what will be proved or disproved by collecting data?

b. What specific data are to be collected? How much is required?

c. What is the source of the data? How much historical data is needed?

d. Who should record the data?

e. Over what time period should the data be collected?

2.2

An example might be to determine why orders are placed late. In this situation the X axis of our check box would be the various stages in the ordering process from producing the requisition, agreeing the tender list, placing the enquiry, answering tenderers' questions, evaluating tenders, and so on until eventually placing orders. On the Y axis would be a list of the various departments. A check mark is then placed in the box for any department that is late with their input to a particular stage of the process. At the bottom of the sheet there would be an indication that the data was collected from time period a to time period b.

2.3

An advantage of check sheets is that specific data such as date, time period, or duration does not need to be recorded. In the above example, just a tick that someone was late.

3   Pareto and Other Diagrams

In general, the following require one to decide how to classify data and the time period covered.

3.1

Line Graphs: These are generally used to present data over a period of time in order to indicate a trend in the characteristic being measured. The classic project example is the use of 'S' curves.

3.2

Pie Charts: These generally show how a fixed amount is distributed or used for different purposes.

3.3

Bar Charts/Graphs: These can either be horizontal or vertical bars, with the units being measured on one axis and time measured on the other axis (usually the X axis).

3.4

Histograms (vertical bars): These show the frequency of actual measurements (on the Y axis) in equal groups of measurements (on the X axis).

3.5

Pareto Diagrams: These are a special form of the vertical bar chart or histogram where the groups of measurements are arranged in descending order from left to right. The exception is the group of measurement classified as 'other' that, if used, is always positioned on the far right of the diagram (even when it is not the smallest of all the groups).

3.5.1

Pareto analysis tends to show that 80 per cent of the problems are caused by 20 per cent of the causes – the 80/20 rule.

4   Cause and Effect – Ishikawa or Fish Bone Diagram

4.1

This is the most complex diagram, but it is fairly straightforward to construct. Draw a horizontal arrow (left to right) with a rectangle at the right-hand end. The effect (problem or issue) is written in the box. See Figure VI.K.1.

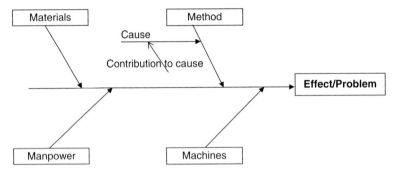

Figure VI.K.1

4.2

The major categories of possible causes are then connected to the line from above and below. As a starting point use the 4 M's: *materials, method* (process), *manpower* (people) and *machines* (equipment). Use other categories, such as environment, if necessary.

4.3

Brainstorm possible causes of the effect and write them on the diagram under the relevant category.

4.4

Using the experience and capability of the team, determine/identify the most likely cause. It may then be necessary to collect and analyse data in order to verify the cause.

5   Force Field Analysis

5.1

Write the subject/problem/issue/proposition to be addressed at the top of a sheet of flip chart paper. Divide the sheet in half with a vertical line. The vertical line represents

PREFABRICATION

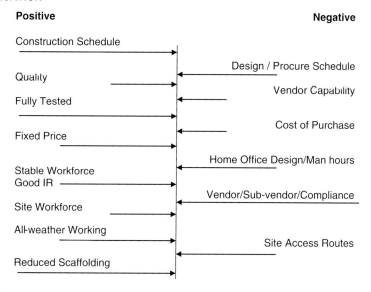

Figure VI.K.2

the present situation and the extreme right hand side of the sheet represents the desired outcome. See Figure VI.K.2.

5.2

The left-hand side is titled 'Positive' for recording the positive forces that help us to achieve our objective. The right-hand side is titled 'Negative' for recording the forces working against our ability to achieve the desired outcome.

5.3

Brainstorm all the issues that are in favour of the proposition and assess/decide whether they are large, medium, or small forces contributing to the desired outcome. Draw large, medium, or small arrows (left to right), and label them accordingly.

5.4

Similarly, brainstorm all the negative forces preventing you from achieving your objective with large, medium, or small arrows from right to left.

5.5

Firstly, select each of the small positive forces in turn. Then get the group to work on how to increase the perceived size of these forces (one could, perhaps, use a cause-and-effect diagram for this process).

5.6

Secondly, select in turn, each of the large negative forces and work on how to reduce/mitigate the issues working against you in achieving your desired outcome.

5.7

If necessary, work on the medium-sized forces in a similar manner.

Section L   Report Writing

"I am sorry to write such a long letter. I did not have time to write a shorter one."

Winston Churchill.

A project report is likely to fall into two broad categories as follows:

a. Recording data and information, such as a progress report
b. Investigation into a specialized topic, such as a problem-solving investigation.

The first impression that the reader will receive will be formed by the appearance of the document. Unless you are constrained by a specific format or house style, you have a choice in the design of the document: the layout, the typefaces, line spacing, and colours used. The writing style will then begin to have an impact when the reader finds the document clear and easy to read. Make sure that any diagrams or illustrations are explained and support the text.

Use special classifications such as *Confidential* sparingly and only when it genuinely applies.

Make sure the report is dated.

1   The Report Objective

1.1

All reports have a purpose, so state the objective clearly right up front. Express the topic clearly and describe the problem. Explain why the issue is important enough to merit a report. Is the report presenting information that requires a decision to be made, or is it requesting approval of the recommendations?

1.2

Choose a clear, complete, and concise title for the report and title page.

2   The Reader

2.1

In the same way that you should consider your audience in making a presentation (see Section G Presentation Skills, subsection 6), you should consider the reader. Project reports are likely to be transmitted to others such as:

a. An external client
b. Senior management
c. For project records

Consequently, nearly all project reports will require a formal style.

2.2

Project reports are most likely to be written by other members of the project team. Further, as the project manager, you are likely to be one of the main recipients of a report into a problem-solving investigation. Consequently, if you have particular requirements, make sure they are incorporated into the project management procedures.

2.3

Any report will consume man hours. Further, none of the recipients of a project report are likely to have much time to read it. Consequently, the report should be concise, well argued, supported by sound analysis, and have recommendations for action.

2.4

The main purpose of a project report is to enable the recipient to make a decision so that the project can move forward effectively. Alternatives need to be discussed, but the conclusions need to be clear and the recommendations unambiguous.

3 The Material for the Report

3.1

Don't forget to include the background material, such as correspondence, minutes, other file notes, and so on.

3.2

Collect the facts and ideas about the subject by reading, observation, and discussion with the project team members.

3.3

Brainstorm the material and data to be used and collate it into similar groupings or natural divisions of the subject. Eliminate material that is not relevant.

3.4

Investigations into technical problems must include enough detail to enable another company specialist to check the findings.

3.5

Check all the information you gather for accuracy.

4   The Report Structure

4.1

Progress reports will have a very clear and consistent structure that has been agreed with the client. See Part IV, Section M concerning progress reporting.

4.2

A typical format is listed below. However, items b and c are sometimes interchanged. This is recommended for a proposal executive summary – you want the client to read it first without having to look for it. They know the contents; they defined them in the enquiry. However, for a normal report, it seems illogical. There is no reason why executives cannot read the contents page and then the summary.

a. Title page with the project name and number, the date of the report, and the title or subject
b. The contents list
c. The executive summary
d. The introduction
e. The main body of the report
f. Conclusions and recommendations
g. References, sources of data and appendices.

5   The Executive Summary

5.1

For a proposal the executive summary should be between 1.5 to 2.5 (3 absolute maximum) pages long. It will start with why the company should be awarded the project.

5.2

A normal report should be 1.5 (or less) to 2 (maximum) pages long. It will have a section briefly outlining the purpose of the report. It will then summarize the main findings and recommendations.

6   Introduction to the Report

6.1

The introduction will outline the purpose of the report in slightly more detail. It will state the approach used in the investigation, together with any other resources used. It is distinguished from the executive summary by not having any findings or conclusions.

7   The Body of the Report

7.1

Describe the procedures used in the investigation to gather the facts, so that someone else could repeat the process if necessary.

7.1.1
Be careful to distinguish between fact and opinion. Check your facts.

7.2

Analyse the situation. State your assumptions and the limitations inbuilt into the investigation.

7.3

Make an outline, linking the various groupings in a logical order. Show how subordinate problems are interrelated. Progress reports will have the information in a sequential order.

7.4

Choose section headings for each main grouping or subject division.

7.5

Discuss advantages and disadvantages of alternatives objectively. State the criteria used to evaluate the alternatives. In a project, this discussion and analysis is bound to involve cost and schedule implications, leading to a natural conclusion.

7.5.1
Identify the consequences of not taking any action.

7.6

Justify the main recommendations.

8   Writing the Report

8.1

Write a very brief statement of what you want to say or achieve with your report, and use this as a reference point when you are writing the main elements of the report.

8.1.1

What type of report are you writing? If it is to get agreement to a specific course of action, you will need to persuade and sell your case. If the report is about a technical problem, it will need to explain and clarify the issues.

8.1.2

Section G, Presentation Skills, identifies the following types of presentations: to inform, to influence, and to express an opinion. The paragraphs 2.2, 3.2, and 4.2 on the body of the presentation may help with your approach to writing the report.

8.1.3

Paragraph 2.1 above makes the point about considering the reader and what switches on the people who are going to read the report. Focus on the primary reader or decision-maker. Subsection 4 of Section E on Personal Skills, addresses different types of people as follows:

a. Paragraph 4.1 for an ideas-orientated person
b. Paragraph 4.2 for a process-orientated person
c. Paragraph 4.3 for a people-orientated person
d. Paragraph 4.4 for an action-orientated person

8.2

Some guidelines to help make the text more effective
a. Use a standard font, readable font size, and a line spacing of 1.15. Make it look good.
b. Write in plain English, using familiar words. Write using short sentences.
c. Do not use abbreviations without first spelling them out in full. Do not use buzzwords. Define all special terms and/or terms used in a special way.
d. Be extremely cautious of using 'cut and pasted' boiler plate from other sources. This can cause serious errors.
    I know of one case where the contractor submitted a proposal to client X, and all the way through one section, it said client Y!
e. Headings should be used to highlight important areas and to help the reader find what they want.
f. Eliminate padding. Practice your précis writing.
g. Avoid vague phrases, and be specific.
h. Look at issues in a positive way, rather than being negative.
i. Work at being consistent in your use of headings, capitalization, and numbering.
j. If you use numbered paragraphs, do not use more than three numbers with two decimal points. Alternate numbers and letters for subsections.

8.3

With today's technology, there is no excuse for spelling mistakes. Use a thesaurus. Use the appropriate English spelling using Z's. Organization is not an Americanism; it is the correct English spelling.

8.4

Use numbers for quantities below 10 and spell them out in full for larger numbers. For sums of money, use numbers and then spell them out so that any mistake is highlighted.

8.5

Consider what use you can make of diagrams or illustrations in order to reduce the amount of descriptive text. Make sure every diagram or illustration has a title, is numbered, and referred to in the text.

8.6

Be sure to reference sources of information and other people's material. This enables the reader to follow up the background material should they so wish. Academic reports usually list the references on a page(s) at the end of the report. I prefer to read them as I read the report and place them at the bottom of the page. This avoids having to flip backwards and forwards, interrupting the reader's flow. If there is a lot of detail, keep it for an appendix.

8.7

Check your sources and facts.

9   Conclusions and Recommendations

9.1

The material in these sections will be in more detail than that presented in the executive summary. Emphasize the significance of the subject matter.

9.2

You may not need both a conclusions and a recommendations section. The conclusions should flow naturally from the body of the report and should be supported by the evidence and data. There should not be any new material in this section.

9.3

In a project report, there will almost certainly be a requirement to finish with some advice on the action to be taken in the recommendations.

9.4

State the cost and schedule implications of the chosen recommendation.

10 Appendices

10.1

If there are a lot of data, rather than clutter up the body of the report or interrupt the flow of the material, place it in an appendix. It may be useful, however, to summarize some of the key information so as to prevent people having to refer to the appendix in order to understand the arguments being put forward.

10.2

If the report contains numerous references to project documents, it may be helpful to compile a separate list as an appendix.

11 Finalizing the Report

11.1

Recheck the contents page and page numbering

11.2

If you have the luxury of some time, leave the report for half a day and then reread it to see if it achieves what it was meant to do. Criticise it objectively as if it were someone else's work. Is the style readable, and does it flow?

11.2.1

Can you improve the executive summary? Senior management will form an opinion of your abilities from items of this nature.

11.3

Is the relationship between the diagrams and the text clear and as explicit as possible?

11.4

End with a punch line, if you have a good one.

Having mastered everything in this book, you will have nothing to do and your projects will be boring! (see Introduction)

Abbreviations

This list contains all of the technical abbreviations used in the book and additional abbreviations that a project manager might come across.

ACE	Association of Cost Engineers
ACWP	Actual Cost of Work Performed
AFC	Approved for Construction
AFD	Approved for Design
AFP	Approved for Purchase
AI	Artificial Intelligence
ALARP	As Low as Reasonably Possible
AOP	Asset Operations Plan
APM	Association for Project Management
AR	Action Required
BAC	Budget at Completion
BATNA	Best Alternative to a Negotiated Agreement
BCWP	Budgeted Cost of Work Performed
BCWS	Budgeted Cost of Work Scheduled
BIM	Building Information Modelling
BMS	Building Management System
BOD	Build Operate Deliver
BOL	Build Operate Lease
BOO	Build Own Operate
BOOM	Build Own Operate Maintain
BOOST	Build Own Operate Subsidize Transfer
BOOT	Build Own Operate Transfer
BOT	Build Operate Transfer
BRT	Build Rent Transfer
BTO	Build Transfer Operate
CA	Competent Authority
CAD	Computer-Aided Design
CAPEX	Capital Cost Estimate
CCGT	Combined Cycle Gas Turbine
CCS	Crown Commercial Service
CD	Compact Disc

Effective Project Management: Guidance and Checklists for Engineering and Construction, First Edition. Garth G.F. Ward.

CDM	Construction Design Management (Regulations)
CEO	Chief Executive Officer
CF	Completer Finisher
CFO	Chief Financial Officer
CFR	Cost and Freight
CH	Chairman
CIC	Construction Industry Council
CIF	Cost Insurance and Freight
CIOB	Chartered Institute of Building
CIP	Carriage and Insurance Paid
CO	Coordinator
COCO	Contractor Owned Contractor Operated
COO	Chief Operating Officer
COQ	Cost of Quality
CPA	Cost Price Adjustment
CPD	Continuing Professional Development
CPI	Cost Performance Index
CPM	Critical Path Method
CPT	Carriage Paid To
CTQ	Cost Time Quality
CTR	Cost Time Resource
CV	Cost Variance
CV	Curriculum Vitae
CW	Company Worker
DAC	Design Acquire Construct
DAF	Delivered at Frontier
DAP	Delivered at Place
DAT	Delivered at Terminal
DBFO	Design Build Finance Operate
DBOM	Design Build Operate Maintain
DBOT	Design Build Operate Transfer
DCF	Discounted Cash Flow
DCMF	Design Construct Manage Finance
DES	Delivered Ex Ship
DEQ	Delivered Ex Quay
DDP	Delivered Duty Paid
DDU	Delivered Duty Unpaid
DUR	Duration
DVD	Digital Video Disc
EBITDA	Earnings before Interest, Taxes, Depreciation, & Amortisation
ECGD	Export Credits Guarantee Department
EEC	European Economic Community
EF	Early Finish
EFD	Engineering Flow Diagram
EHS	Environment Health and Safety
EIA	Environmental Impact Assessment
EIS	Environmental Impact Statement
EOI	Expression of Interest

EOY	End of Year
E&P	Exploration and Production
EPC	Engineering Procurement and Construction
EPCC	Engineering Procurement Construction and Commissioning
EPCI	Engineering Procurement Construction and Installation
EPCM	Engineering Procurement and Construction Management
EPIC	Engineering Procurement Installation and Commissioning
ERM	Enterprise Risk Management
ES	Environmental Statement
ES	Early Start
EU	European Union
EV	Enterprise Value
EV(M)	Earned Value, EV Management
EXW	Ex Works
FAS	Free Alongside Ship
FAX	Facsimile
FBOOT	Finance Build Own Operate Transfer
FCA	Free Carrier
FEED	Front End Engineering Design
FEL	Front End Loading
FF	Finish to Finish
FF&E	Furniture Fittings and Equipment
FID	Final Investment Decision
FM	Facilities Management
FOB	Free On Board
FTP	File Transfer Protocol
GM	General Manager
GM	Gross Margin
GOCO	Government Owned Contractor Operated
GRASP	Global Risk Assessment and Strategic Planning
HAZAN	Hazard Analysis
HAZID	Hazard Identification (study)
HAZOP	Hazard and Operability (review/study)
HO	Home Office/Head Office
HS&E	Health Safety and Environment
HSE	Health and Safety Executive
HSSE	Health Safety Security Environment
HUC	Hook Up and Commissioning
HVAC(&R)	Heating, Ventilating and Air Conditioning (and Refrigeration)
IBRD	The International Bank for Reconstruction and Development
ICB	International Competitive Bidding
ICC	International Chamber of Commerce
ICE	Institution of Civil Engineers
IChemE	Institution of Chemical Engineers
IDA	International Development Association
IEE	Institution of Electrical Engineers
IIM	Importance or Interest/Influence Matrix

IM	Implementer
IM	Information Management
IMechE	Institution of Mechanical Engineers
INCOterms®	International Commercial Terms
IoD	Institute of Directors
IPA	Infrastructure and Projects Authority
IPMA	International Project Management Association
IPR	Intellectual Property Rights
IR	Industrial Relations
IRM	Inspection Repair Maintenance
IRR	Internal Rate of Return
IS	Information Services
ISO	International Organization for Standardization
ISBN	International Standard Book Number
IT	Information Technology
ITB	Invitation to Bid
ITP	Instruction/Intention to Proceed
ITPD	Invitation to Participate Document
ITT	Invitation to Treat/Tender
ITT	Instructions to Tenderers
JIT	Just In Time
JV	Joint Venture
KPI	Key Performance Indicator
LDA	Large Data Analysis
LF	Late Finish
LfE	Learning from Experience
LIM	Like, Intend, Must
LS	Lump Sum
LS	Late Start
LSTK	Lump Sum Turnkey
MBTI	Myers Briggs Test Instrument
MBWA	Management by Wandering or Walking Around
MDR	Master Document Register
ME	Monitor Evaluator
MMT	Mobile Maintenance Train
MOM	Minutes of Meeting
MoSCoW	Must haves, Should haves, Could haves, and Won't haves
MPA	Major Projects Association
MTO	Material Take Off
MVP	Minimum Viable Product
NA	Network Analysis
NAO	National Audit Office
NB	Nota Bene
NDT	Non-destructive Testing
NEC	New Engineering Contract
NEDO	National Economic Development Office

NGO	Non-Governmental Organization
NICE	Needs Interests Concerns Expectations
NPV	Net Present Value
OB	Organizational Behaviour
OBS	Organization Breakdown Structure
OEM	Original Equipment Manufacturer
OGC	Office of Government Commerce
OH&P	Overhead and Profit
OJEU	Official Journal of the European Union
OM	Operations Manager
O&M	Operations and Maintenance
OPEX	Operating Cost Estimate
OSBL	Outside Battery Limits
OS&U	Offsites and Utilities
P3M	Project, Programme, and Portfolio Management
PC	Personal Computer
P&ID	Process and Instrument Diagram
PBS	Product Breakdown Structure
P&WBS	Product Breakdown Structure and Work Breakdown Structure
PCT	Project Change Triangle
PEP	Project Execution Plan
PFI	Private Finance Initiative
PfM	Portfolio Management
PHSERS	Project Health Safety and Environment Reviews
PI	Public Indemnity (insurance)
PI	Profitability Index
PIP	Process Industry Practices
PL	Plant
PL	Plain Line
P&L	Profit and Loss
PM	Project Manager
PMC	Planning, Monitoring, and Control
PMI	Project Management Institute
PMO	Project Management Office
PPM	Project Portfolio Management
PPP	Public Private Partnership
PQQ	Prequalification Questionnaire
PR	Public Relations
PSP	Project Selection and Prioritisation Process
QA	Quality Assurance
QC	Quality Control
QMS	Quality Management System
QRA	Quantified Risk Assessment
QRA	Quantitative Risk Analysis
R&D	Research and Development
RACI	Responsible, Accountable, Consulted, Informed
RBS	Risk Breakdown Structure

RFI	Request for Information
RFP	Request for a Proposal
RFQ	Request for a Quotation
RI	Resource Investigator
RIBA	Royal Institute of British Architects
RICS	Royal Institution of Chartered Surveyors
ROI	Return on Investment
ROV	Remote Operated Vehicle
ROW	Right of Way
RR	Response Required
RTFC	Read the Full Contract
S&C	Switch and Crossing
SH	Shaper
SMART	Specific Measurable Achievable Realistic/Relevant Time related
SMARTER	SMART plus Evaluated and Reviewed
SME	Small and Medium-sized Enterprises
SIG	Specific Interest Group
SOR	Statement of Requirements
SP	Specialist
SPE	Society of Petroleum Engineers
SPI	Schedule Performance Index
SRA	Schedule Risk Analysis
SS	Start to Start
STOR	Short-Term Operating Reserve
SV	Schedule Variance
SWOT	Strength Weakness Opportunities Threats
TBM	Tunnel Boring Machine
TCPI	To Complete Performance Index
TF	Total Float
TLA	Three Letter Acronym
TOR	Terms of Reference
TPM	Total Preventative Maintenance
TQM	Total Quality Management
TW	Team worker
USB	Universal Serial Bus
USP	Unique Selling Point
VA	Value Analysis
VAT	Value-Added Tax
VE	Value Engineering
VM	Value Management
WACC	Weighted Average Cost of Capital
WBS	Work Breakdown Structure

Index

Effective Project Management: Guidance and Checklists for Engineering and Construction,
First Edition. Garth G.F. Ward.
© 2018 John Wiley & Sons Ltd. Published 2018 by John Wiley & Sons Ltd.